AF885419

SOTHIS PRESS

Sothis Press aims to completely reimagine the public domain book with an emphasis on universal access to captivatingly designed print editions. We believe that technology will enable every single book ever published to be perpetually accessible in beautiful print form. Sothis is beauty in ubiquity.

www.sothis.press
1 Waterhouse Square, London, EC1N 2ST

Problems in Physics

Derived From Military Situations and Experience

United States War Department

SOTHIS PRESS

ISBN 978-1-0365-0430-4

This edition is published by
Sothis Press © 2023. The company
is registered in England and Wales.
The Company Number is 14900389.

PROBLEMS IN PHYSICS

DERIVED FROM MILITARY SITUATIONS
AND EXPERIENCE

WAR DEPARTMENT
COMMITTEE ON EDUCATION AND SPECIAL TRAINING
WASHINGTON

PREFACE

One of the lessons taught by the intensive war training experience was that rapid progress in learning can be made when men are trained for definite jobs. Hence a better result will be secured from the technical class-room instruction for the Reserve Officers' Training Corps if the product needed by the army is first defined in terms of the jobs and operations which an officer will be called upon to perform.

Such definitions will also assist college teachers in their regular work by supplying problems and other material which will enable them to connect their instruction with the broad field of application in which the mastery of scientific principles is objectively revealed. Instruction based on problems and job sheets of graded difficulty seem to offer as effective a means as can be found of developing the ability to do things in civilian life as well as in the army.

The following collection of problems in several important branches of Physics has been made by Dr. Homer L. Dodge, head of Department of Physics of the University of Oklahoma, with the hearty cooperation of the Liaison Officers and officers of the various corps. It is one of a series on various technical subjects being prepared by the War Department for the purpose of helping teachers who give technical instruction to the students of the R. O. T. C. Those who can solve these problems will be well qualified in these subjects for service in the several staff corps and for civilian pursuits in which the subjects treated are involved.

C. R. MANN,
Chairman, Advisory Board.

TABLE OF CONTENTS

ILLUSTRATIONS

PART I.

PART II.

PART III.

PART IV.

PART V

PART VI.

PART VII.

PART I.

THE 3-INCH TRENCH MORTAR

THE 3-INCH TRENCH MORTAR

One of the most interesting pieces of artillery developed to meet the conditions of the World War is the trench mortar. A short range gun was needed which would be portable, easily handled and cheap to manufacture. All of these requirements were met in the trench mortar.

The 3-inch mortar is shown in action in Figure 1. It consists of a four foot length of heavy seamless steel tubing forming the barrel, a bipod to support the barrel, and a base plate to distribute the shock of recoil over a sufficient surface. The bipod, which carries the traversing and elevating screws for aiming, is clearly visible in the picture but the base plate is concealed by the sand bags piled about the base of the mortar.

The method of firing is unique. All that is necessary is to drop the shell into the mortar. By the time the loader's hands are removed, the projectile comes flying out of the mortar on its way to the distant target. This simple firing operation is made possible by the construction of the shell. Screwed into its base is a short length of steel tubing, or cartridge container, which holds a cartridge very much like a 12 gauge shot-gun shell. When the shell is dropped into the barrel of the mortar it slides smoothly down the bore until the primer of the cartridge is fired as it strikes a firing pin in the base of the mortar. The gases from the cartridge escape through ports in the container into the chamber of the mortar, thus propelling the shell toward the target, the exploded cartridge being carried with it.

To increase the range of the mortar ballistite rings are used, one, two, or three rings, according to the range desired. These propellant charges consist of doughnut shaped silk bags, each containing 110 grains of ballastite powder. They are slipped over the cartridge container and are ignited by the flash of the cartridge which reaches them through the ports in the cartridge container.

A very clear idea of the shell is gained from Figure 1, in which can be distinguished the fuse at one end of the casing and the cartridge container, carrying a single powder ring, at the other.

THE TRAJECTORY

So great is the effect of air resistance upon the trajectory of a projectile that range tables derived from actual experience at the proving ground must be used as a basis for aiming any piece of artillery. It is, however, necessary for any officer whose work involves a knowledge of exterior ballistics, to be thoroughly familiar with the conditions as they would exist were the projectile moving in a vacuum.

The 3-inch trench mortar is particularly suitable as a basis for problems on trajectories in vacuo because of its short range and the correspondingly low velocity of the shell. Air resistance is a far less important consideration than in the case of guns of higher muzzle velocities. Consequently the results obtained from computations based on the assumption of vacuum conditions retain a considerable element of reality.

Nevertheless the effect of air resistance is by no means negligible and an excellent opportunity is afforded for a comparison of actual experience with that which would result from vacuum conditions. In order to make this possible, an abridged Range Table, based on tests at the Aberdeen Proving Ground, is provided on page 28.

PROBLEMS

1. A 3-inch trench mortar is trained upon a machine-gun emplacement 300 yards away. The elevation is 45 degrees and the azimuth is correct. How near will the shell come to the target if the charge consists of a cartridge alone? Compute for vacuum conditions and compare your results with the actual range shown in the range table.
2. The elevation of the mortar is changed to 59 degrees. Compute the distance by which the shell will miss the target.
3. Find by computation whether a hit would be scored with an elevation of 45 degrees if the emplacement were 100 feet above the level of the mortar.
4. What is the maximum range in vacuo that can be secured with the cartridge alone? What is the elevation?

Figure 1. 3-Inch Trench Mortar in Action.

5. An enemy trench is 400 yards away. Can it be successfully shelled using a charge consisting of a cartridge and one powder ring? What will be the angle of elevation (in vacuo)? Compare with the actual range and elevation.
6. When soldiers are instructed to throw hand grenades at the angle which gives maximum range, what angle is specified?
7. An enemy tank is observed 400 yards away. It is advancing at a speed of 6 miles per hour. Assume a muzzle velocity of 259 feet per second and vacuum conditions.

    (a) What will be the time of flight of the shell?

    (b) How far will the tank have moved during the flight of the shell?

    (c) What is the correct angle of elevation?
8. An enemy tank is observed 300 yards away. It is making off at a speed of 6 miles per hour in a direction perpendicular to the line of sight. At what angle in advance of the tank must the mortar be aimed? Assume a muzzle velocity of 180 feet per second and neglect air resistance.
9. An enemy tank is observed 400 yards away. It is making off at a speed of 6 miles per hour in a direction 45 degrees from the line of sight. What is the correct angle of elevation (in vacuo) for the mortar? By what angle (azimuth) should the tank be "led"? Assume a muzzle velocity of 259 feet per second.
10. An enemy tank is observed at a distance of 700 yards. Its speed is 8 miles per hour.

    (a) Compute the allowance which must be made for the distance it covers during the time of flight if you are using a zone three charge. If you are using a zone two charge. Neglect air resistance.

    (b) If the tank is advancing toward the mortar, which charge requires the greater correction of the elevation in order to allow for the distance covered by the tank during the flight of the shell?

    (c) If the tank is moving at right angles to the line of sight, which charge requires the greater change of azimuth?

11. Some trench mortars are assigned the task of keeping the enemy from coming out of a dugout 600 yards distant. The object is to drop shells as nearly vertically as possible into the trench. Is there more than one angle at which these shells can be dropped into the trench, assuming vacuum conditions? If so, compute the angle and state which charge and elevation should be used to secure the desired result.
12. Discuss the relation between the angle of elevation of the mortar and the angle at which a shell strikes. How does air resistance affect this relation?
13. The British used a time fuse on their trench mortar shells. If a time fuse were to be used on our 3-inch trench mortar shells how long a maximum time would have to be provided for, assuming vacuum conditions?
14. What keeps a shell in the air?
15. Report upon the effect of air resistance upon the trajectory of a 3-inch trench mortar shell when fired with the cartridge only.

    (a) What is the formula for range in vacuo with which computations can be made most readily?

    (b) Find the actual reductions of range, caused by air resistance, for the different elevations listed in the Range Table.

    (c) What is the average reduction figured on a percentage basis?
16. Compare the effect of air resistance at different velocities by comparing the average percentage reduction of range for each of the four charges. Compute for 45 degrees, 60 degrees, and 75 degrees elevations only.
17. The enemy trench is at such a range that it will be reached with an elevation of 50 degrees and a zone two charge. If the gunner makes an error of one degree in his elevation by how much will he miss the target? Compute for vacuum conditions and compare result with actual facts as indicated by the Range Table.
18. The enemy trench is at such a range that it will be reached with an elevation of 75 degrees and a zone two charge. If

the gunner makes an error of one degree in his elevation by how much will he miss the target? Compute for vacuum conditions and compare with actual facts as indicated by the Range Table.

19. Explain briefly why it is that a given error in the elevation setting makes more difference in the range the greater the elevation. Will this be true for guns firing at low angles? Why?

20. Does the possibility of a shell being blown back on one's own lines have to be considered?

(a) How high a wind would be necessary with the smallest charge and highest angle of elevation?

(b) With the largest charge and highest angle of elevation?

21. What charges and what angles of elevation could be used in a head wind of 50 miles an hour?

22. You are an officer in the trench warfare section. Your superior officer wishes some material which will show in a striking manner the effect of air resistance.

(a) Prepare material which will show that air resistance increases rapidly with increase of velocity. Use the Range Table and answers to previous problems as much as possible.

(b) Show that the percentage decrease of range due to air resistance is approximately the same for all elevations with the same charge.

(c) Show how the percentage decrease of range due to air resistance increases with range.

(d) Prepare a sketch showing the actual trajectories for minimum and maximum range for each charge compared with the trajectories which would be secured were there no air resistance.

23. An enemy trench is 200 yards distant and 50 feet above the level of the mortar. Compute the angle of elevation (in vacuo) which should be employed.

24. A 3-inch trench mortar is set up 220 yards from an enemy machine-gun emplacement which is 100 feet below. Compute the angle of elevation (in vacuo) which should be employed.

THE 3-INCH TRENCH MORTAR FUSE

As the shell of the 3-inch trench mortar is of the fragmentation type a fuse mechanism to detonate the bursting charge is necessary. Moreover, as the mortar is not rifled and the shell has no guide vanes to give it true flight, it tumbles and may hit the target in any position. To meet these conditions an "all-way" percussion type fuse has been developed. As will be seen from Figure 2, the fuse is by no means a simple mechanism. And yet it is not more complicated than is necessary, for each part has a very defiinte function to perform.

A fuse must do two things—it must arm and it must fire. If the firing mechanism of a fuse were "armed" previous to the firing of the gun there would be grave danger from accidental firing during shipping and handling and, in the case of some mechanisms, detonation would be certain to occur in the bore. Consequently the arming mechanism is a very essential part of a fuse.

The 3-inch trench mortar fuse is armed as a result of the enormous "setback" or inertia reaction accompanying the sudden increase in velocity when the mortar is fired. In this connection it is to be remembered that from a standing start, so to speak, the projectile takes on a velocity of 180 to 372 feet per second, or several times the speed of an ordinary railway train, in a very small fraction of a second. The simplest way of describing the magnitude of the setback stresses is to state the setback as so many pounds force per grain of weight. For example, if a loaded 3-inch trench mortar fuse weighs 1 pound 3 ounces, and the setback is ¼ pound per grain, the fuse will bottom down against the shell casing with a total force of over 2,000 pounds, every part of the fuse contributing its share of the setback force.

When the mortar is fired, the setback pellet compresses the setback pellet spring and releases the safety fork which is pushed out against the barrel of the mortar by the safety fork spring and left behind as the shell leaves the mortar. The safety fork normally extends through the hole in the sleeve of the firing mechanism and prevents the striker from firing the percussion cap. Otherwise the setback of the striker would fire the cap and the shell would burst in the mortar. On leaving the mortar the shell is armed, i. e., in condition to be fired on impact.

During shipment the fuse is protected from arming by a safety pin which extends through the body of the fuse and engages the setback pellet holding it securely in place. Since the safety pin must be removed before loading and might at any time be accidentally removed, the arming mechanism proper must be so designed as to be safe under rough handling and accidental shocks. For this reason the mechanism must not operate too easily. On the other hand the fuse must be sure to arm on leaving the mortar and this will not be assured if the mechanism works with too great difficulty.

These considerations have had to be kept in mind by the designer of the fuse. The mass of the setback pellet and the strength of the setback pellet spring must be determined with reference not only to the normal accelerations experienced in the mortar, but also to the accidental accelerations resulting from such shocks as the shell might experience before and during loading. This nice balancing of the different factors depends upon the successful application of elementary physical principles, especially those involving force, mass, and acceleration and the elasticity of springs.

Note: Arrangements have been made so that sectionalized fuses can be obtained from the Ordnance Department of the United States Army. Requests from institutions where there are Reserve Officers' Training Corps Units should be made through the Commandant. All requests should be addressed to the office of the Chief of Ordnance, United States Army Washington, D. C.: Attention of the Training Section of the Administrative Division.

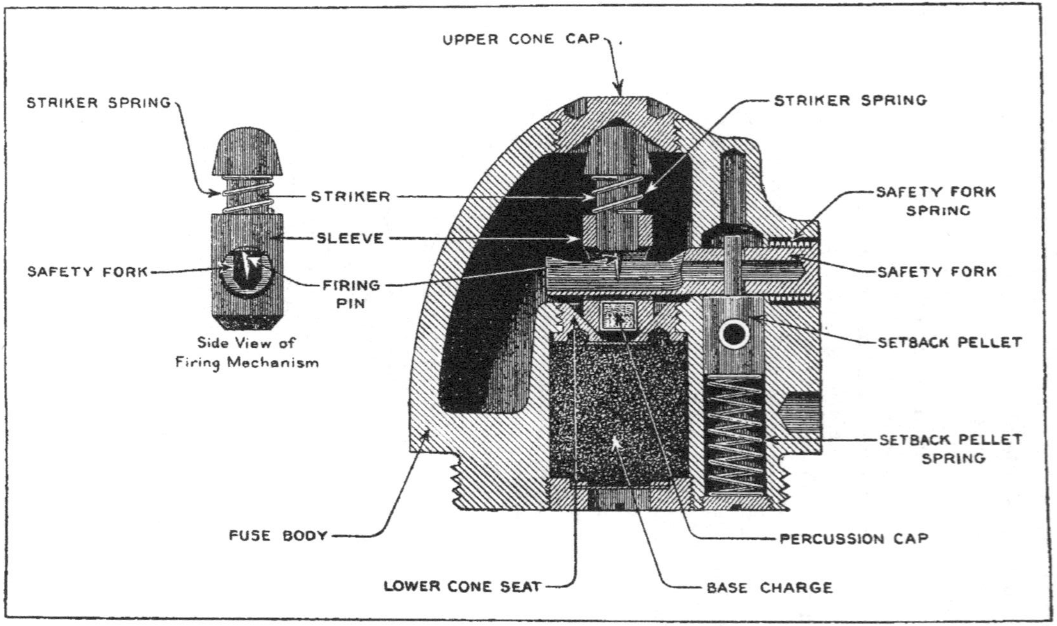

Figure 2. 3-Inch Trench Mortar Fuse, Actual Size

PROBLEMS ON THE ARMING OF THE FUSE

25. Is the fuse sure to arm?[1]

Solution A

(a) The setback pellet must compress the setback spring 3/8 inch in order for the fuse to arm. With how much force must the setback pellet press down upon the spring to arm the fuse? (Study the specifications for the setback pellet spring and determine the force necessary to compress the spring to a height 3/8 inch less than the assembled height of 3/4 inch.)

(b) What is the immediate cause of the force exerted upon the setback spring by the setback pellet when the mortar is fired?

(c) Is the force proportional to the velocity of the shell? Is it proportional to the acceleration of the shell?

(d) Is the weight of the pellet an appreciable part of the force?

(e) Does the force depend upon the mass of the pellet?

(f) Why is it not strictly correct to say that the force depends upon the weight of the pellet?

(g) What must be the acceleration of the shell in feet per second per second in order to insure arming?

(h) What is the physical law which you apply in making the above calculation?

(i) What causes the acceleration of the shell?

(j) Why does the maximum acceleration of the shell occur at the instant of maximum powder pressure?

(k) What relation exists between the acceleration of the shell at any instant and the total force exerted by the powder on the shell at that instant?

[1]The problem of whether or not the fuse will arm can be attacked in two ways. One can examine the fuse and discover how far the setback pellet must compress the spring, and working through the proper steps, arrive at the necessary powder pressure and compare it with the known powder pressure, or he can start with the powder pressure and find the force with which the setback pellet bottoms down on the spring and compare this with the force necessary for arming. The first method is followed in Solution A, the second is Solution B.

(l) What must be the force acting upon the shell in order for the fuse to arm?

(m) For how long a time must this force act?

(n) Must the fuse have any certain velocity before it can arm?

(o) What is the powder pressure, in pounds per square inch, necessary to arm the fuse?

(p) How does this compare with the maximum powder pressure of the weakest charge, which is about 800 pounds per square inch?

Solution B

(a) What is the maximum force acting upon the shell with the weakest charge? Assume the maximum powder pressure, with the cartridge only, to be 800 pounds per square inch.

(b) Does the maximum acceleration of the shell occur at the instant of maximum powder pressure? Explain.

(c) What is the relation between the force acting on the shell and the acceleration of the shell?

(d) What is the maximum acceleration, in feet per second per second, experienced by the shell?

(e) With how much force (pounds) is each grain (weight) of the shell thrust forward?

(f) By what means is this propelling force communicated to the fuse?

(g) With how much force does the fuse "bottom down" against the shell casing?

(h) By what means is the propelling force communicated to the setback pellet?

(i) How much force does the setback pellet spring exert upon the setback pellet normally? How much force can it exert? How much does it exert at the moment of arming? A movement of the pellet of 3/8 inch will cause arming.

(j) Why does the setback pellet not remain in its normal position when the mortar is fired?

(k) With how much force does the setback pellet "bottom down" against the spring?

(l) How much greater is this than is necessary for arming?

26. As far as the arming mechanism is concerned could the 3-inch trench mortar fuse be used with a four inch shell weighing 20 pounds if the maximum powder pressure were 1,000 pounds per square inch?

27. What is the setback in pounds per grain with the maximum charge (cartridge and three rings) assuming a maximum powder pressure of 3,500 pounds per square inch? How does this compare with the propelling force in pounds (force) per grain (weight)?

28. What is the setback required for the arming of the 3-inch trench mortar fuse? How does this compare with the setback produced by the smallest charge, which can be assumed to produce a maximum powder pressure of 800 pounds per square inch?

29. You are preparing the specifications for a setback pellet spring which will permit arming with a setback of 50 per cent of that normally experienced with the smallest charge. Assume the maximum powder pressure of the smallest charge to be 800 pounds per square inch, and that the free height of the spring is to be 1.5 inches. Specify the force with which the spring must compress to an approximate height of .5 inch.

30. How much more difficult to arm would the mechanism worked out in Problem 29 be than that of the regular fuse? What possible advantages would there be in the proposed form?

31. If the loader should accidentally drop the shell, with the safety pin removed, from a height of three feet, head down, into soft ground into which it penetrates two inches, would it be safe for him to use the shell? Find the height from which it could fall without arming.

32. If the loader accidentally drops the shell, with the safety pin removed, base down on a rock is it safe for him to use the shell? Find the height from which it could fall without arming.
33. Study the drawing of the fuse and the specifications for the springs and find the force tending to expel the safety fork from the fuse body. To what value has this been reduced at the time the safety fork is just disengaging the sleeve?
34. Why does friction play a more important part in restraining the motion of the safety fork than it does in the case of the set-back pellet?

THE FIRING OF THE FUSE

The firing mechanism of the 3-inch trench mortar fuse is of the "all-ways" type, i. e., it will operate no matter in what position the shell strikes.

Figure 2 shows the construction and operation of the firing devices. The movable parts, a striker and a sleeve, are supported between the conical surfaces of the upper cone cap and the lower cone seat. If the shell hits head on, the momentum of the sleeve carries it forward and brings the cap in the base of the sleeve against the firing pin of the striker. If the shell hits tail on, the striker hits the cap. If the impact is on the side the two will be brought together through the action of the conical faces of cap and seat. No matter on what position the shell strikes, the fuse is sure to fire.

The percussion cap flashes through a hole in the lower cone seat and ignites the base charge. The base charge, acting through the appropriate detonator and booster charges, carried in the head of the shell proper, detonates the bursting charge and explodes the shell.

PROBLEMS

35. If a shell with a velocity of 300 feet per second strikes head on in ground into which a dummy would penetrate six inches, will it burst?

(a) Why is it necessary to assume a depth of penetration?

(b) What is the average acceleration in feet per second per second?

(c) How much is the average resistance (pounds force) of the ground?

(d) What force does the shell exert upon the ground?

(e) With what force in pounds (force) per grain (weight) do the various parts of the fuse and shell press forward?

(f) What is the approximate force with which the striker hits the cap?[1]

(g) Justify your answer to the main question.

36. The shell hits the ground, tail on, with a velocity of 200 feet per second. A dummy would penetrate six inches. Justify your conclusion as to whether or not the fuse acts. In solving this problem find the resistance of the ground by equating the kinetic energy of the shell against the work done by the ground in stopping it. Is this approach easier than that of problem 35?

37. If the shell falls side on into soft ground into which a dummy would penetrate to a depth of three inches, how much of a blow will the cap receive?[1]

38. The shell to be effective must burst immediately on impact. It must not penetrate the ground any appreciable distance. How far will it penetrate before the firing mechanism acts under each of the conditions mentioned below?

(a) Assume dummy penetration of one foot with head on impact at 750 feet per second.

(b) Assume dummy penetration of six inches with side on impact at 200 feet per second.

(c) Assume dummy penetration of two feet with tail on impact at 300 feet per second.

39. If the fuse should be accidentally armed, will the shell explode when dropped 2.5 feet, tail down, on a concrete foundation? When dropped 3.5 feet?

---

[1]Two factors which do not lend themselves easily to computation affect the actual force. The fact that the firing pin penetrates the percussion cap means that it is not stopped as rapidly as is assumed. This is more than balanced by the fact that the striker spring normally holds the firing pin at a little distance from the cap. This gives the striker room to acquire (relative) velocity and so strike a smarter blow. Consider your experience when standing in a trolley car when it suddenly stops. Do you prefer to come up against some solid part of the car or to let someone else act as a buffer? If you must use something solid to stop you, do you prefer to be close to it or already leaning against it or to come at it from a distance of a foot or two?

40. If the fuse should be accidentally armed, will the shell explode when dropped 3.5 feet, tail on, into soft ground into which it penetrates 2 inches?

41. If the fuse should be accidentally armed during or before loading will it fire in the barrel of the mortar? If so, at what point on the barrel, approximately?

THE FUSE TESTS

Even though the various parts of the fuse are made to specifications and are subjected to rigid inspection and tests it is found desirable to take samples from each lot of assembled fuses and subject them to final tests.

A—TESTS INVOLVING THE ARMING MECHANISM

Each fuse to be tested is housed in a suitable container and dropped upon an iron anvil.

The Arming Test: The fuse is dropped six feet, base down, with the safety pin removed. It must arm but must not fire.

The Safety Test: The fuse is dropped three feet, base down, with the safety pin removed. It must not arm.

PROBLEMS

42. (a) How much more severe could the safety test be made? Compute the maximum possible drop.

(b) Why is it desirable to have the safety test considerably less severe than the limit to which it could be carried?

43. (a) How much more severe could the arming test be made? Compute the minimum possible drop.

(b) Why is it desirable to have the arming test, considerably less severe than the limit to which it could be carried?

44. Find, if possible, from the conditions of the arming test the smallest powder pressure with which the fuse will be sure to arm.

45. Show that the movement of the setback pellet can be entirely neglected in the computations employed in discovering whether the fuse will arm as a result of the accelerations experienced in firing. Show that the movement of the pellet is of the greatest importance in finding the accidental drops which a fuse will stand.

B—TESTS OF THE FIRING MECHANISM

Firing Test: Dropping for the firing test must be done with the fuse in various positions such as normally would be encountered in the field. The percussion cap or primer must fire when armed fuses are dropped from a height of three feet upon an iron anvil.

Percussion Caps: The testing machine is fitted with a four-ounce drop weight with a firing pin of the same shape as the one used in the fuse. All caps tested must fire under the blow delivered when the weight and pin drop from a height not exceeding eight inches.

PROBLEMS

46. Study the firing tests and the specifications of the striker spring. Is the strength of the striker spring an appreciable consideration in the firing tests? Consider for head on, tail on, and side on hits.
47. Show conclusively that the force exerted by the striker spring is not an appreciable consideration in determining whether or not the fuse will fire under such conditions as are actually met. Assume that a compression to solid height is necessary to set off the cap.
48. Which is the more severe test, the firing test or the cap test? How much more severe? In answering these two questions do you neglect the striker spring? If so, explain why this is permissible.
49. Compare the firing test with the firing of the shell in actual warfare. Is the firing test too severe? If you think the firing test should be more severe give your reasons.
50. Imagine yourself called upon to make a report concerning the suitability of the percussion cap test. Such a report would necessitate a study of the relation between this test and the weakest blow with which the cap would actually be struck by the firing mechanism. Make such a comparison and state whether you believe the test to be sufficiently severe.

ABRIDGED RANGE TABLE (YARDS)
3-INCH TRENCH MORTAR

Elevation in Degrees	ZONE			
	First	Second	Third	Fourth
40	295	481	632	747
45	291	472	622	732
50	280	455	602	707
55	265	430	564	666
60	243	396	512	608
65	216	354	453	536
70	186	306	388	451
75	152	238	314	357
Muzzle Velocity feet per second	180	259	318	372
Charge	Cartridge only	Cartridge and one ring	Cartridge and two rings	Cartridge and three rings

SPECIFICATIONS FOR SPRINGS

Setback Pellet Springs must compress to a height of .232 inches with a weight of not less than 28 ounces nor more than 35 ounces. They must return to a free height of not less than 1.35 inches after compression. In problems assume proper compression with 30 ounces and free height of 1.35 inches. The assembled height is ¾ inch.

Safety Fork Spring: Same specifications as for Setback Pellet Spring.

The Striker Spring consists of three turns. It has a free height of .28 inches and must compress to a solid height of .06 inches with a dead weight of 6 to 8 ounces. In problems assume 8 ounces.

MISCELLANEOUS DATA
(3-INCH TRENCH MORTAR)

Length of barrel	4	feet
Distance shell travels in bore	3.5	feet
Weight of shell, fully loaded	11 lbs.	11 oz.
" " fuse	1 "	3 "
" " setback pellet	.212	ounces
" " striker	.247	ounces
" " sleeve	.297	ounces
" " safety fork	.317	ounces
" " setback pellet spring	.017	ounces
" " striker spring	.0035	ounces

PART II.

BLOCKS AND TACKLE

BLOCKS AND TACKLE

The accompanying sketches show various tackle riggings with typical applications. In practice there is usually a certain limited amount of equipment and the problem is to make the best use of what is at hand.

Among the facts which govern the effective handling of a given job are the following:

1. The strength of the ropes or cables must not be exceeded.
2. The rigging should be arranged so as to distribute the load advantageously between lengths of heavy and light rope.
3. Available force must be applied to the best practical advantage.
4. As much distance as possible should be covered without "overhauling." Under certain conditions the full distance must be covered.
5. Speed rather than mechanical advantage is often required.
6. Space limitations must be considered.
7. Ropes and cables should not be cut.

In determining the mechanical advantage it is convenient to make a rough sketch and mark the block hooks and each return with a symbol or number to indicate the force or tension. Usually the number is used which represents the ratio of the force in question to the applied force.

If friction is considered, the tension in the fall should be reduced 10 per cent at each sheave or, in the case of two-block tackles, a reduction of mechanical advantage of 5 per cent for each sheave may be assumed. These figures are, of course, only approximate. In actual practice, the applied force is often sufficiently variable to take care of the friction.

A horse will pull from 300 to 400 pounds for a short time on a horizontal lead line, a man 40 to 60 pounds. A man can exert a force of 16 pounds continuously on the handle of a winch and double this for short periods.

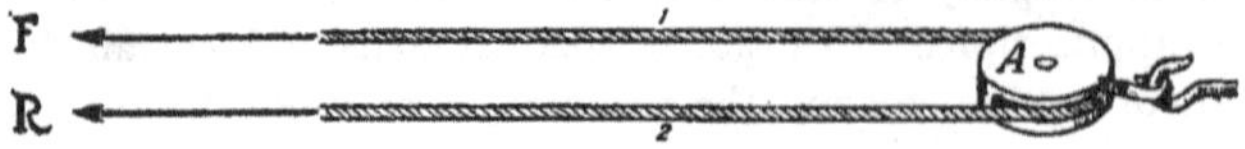

Single Whip

SINGLE WHIP

1. How many men would be required to move the log in Figure 1 with a direct pull, if the six men are handling it successfully as shown in Figure 1?
2. Why does the nature of the footing to the right of the tree in Figure 1, determine whether or not the Single Whip is of advantage?

Running Tackle

RUNNING TACKLE

3. What is the practical utility of using the single block as a Running Tackle as in Figure 2?
4. What difficulties are introduced into the situation represented in Figure 2 if immediately behind the tree there is a stream into which the log has to be launched? How would you meet these difficulties?
5. What are the relative mechanical advantages of the Single Whip and the Running Tackle?
6. Which needs the stronger rope, the Single Whip or the Running Tackle? Which the stronger tree?

Gun Tackle

GUN TACKLE

7. What gain in mechanical advantage over the Single Whip is secured by introducing the second block as shown in the Gun Tackle of Figure 3? What advantage over the Running Tackle? What loss is there in speed?

8. What gain in mechanical advantage over the Single Whip is secured by introducing the second block as shown in the Gun Tackle of Figure 4? What advantage over the Running Tackle? What loss is there in speed?

9. Compare the two Gun Tackles on the basis of mechanical advantage and speed.

10. The kegs in Figure 3 weigh 120 pounds each. Consider friction and find: (a) with what force the man is pulling, (b) the tension in each return.

11. The kegs shown in Figure 3 weigh 120 pounds each. The rope on which the man is pulling is at an angle of 45 degrees. What is the stress in the rope supporting the upper

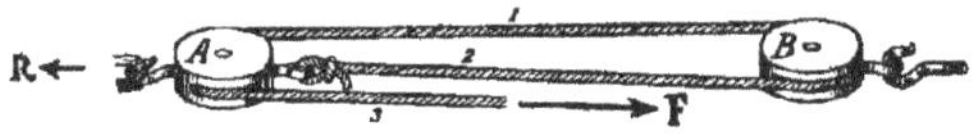

Gun Tackle

block? What is the force acting on the hook of the lower block? Neglect friction and weights of parts.

12. Suggest a desirable change in the rigging of the Gun Tackle shown in Figure 4, in case the tree is in a swampy place and there is no good footing for twenty or thirty feet on the side toward the I beam. How much harder will the men have to pull with the new rigging?

13. The men shown in Figure 4 are each pulling with a force of 50 pounds. Taking account of friction find; (a) the force exerted on the I beam, (b) the force exerted on the tree, (c) the tension in each return.

14. How many men would be required to drag an 800 pound steel I beam across a cement foundation, using the Gun Tackle? Look up necessary data.

15. Some 40 pound kegs are to be raised from a flat car to the third floor of a building. Would you use the Whip on Whip or the Gun Tackle? Why?

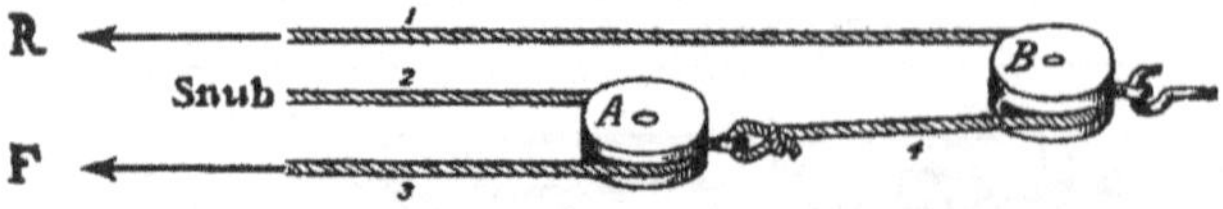

Whip on Whip

WHIP ON WHIP

16. What is the mechanical advantage of the Whip on Whip?
17. What can you accomplish with the Whip on Whip which you cannot accomplish with the Running Tackle?
18. If you are going to use a horse to move a load requiring a force of 800 pounds and have available plenty of rope which will stand 300 pounds, will you use the Whip on Whip, the Running Tackle, or the Gun Tackle? Which will you use if the resistance is 500 pounds? 200 pounds?
19. It requires a force of 1,000 pounds to drag a gun 60 feet to the base of a high wall in which there is a stay bolt conveniently located. There is available 75 feet of heavy rope capable of standing a load of 1,500 pounds and 60 feet of rope which will stand 600 pounds. Sketch the rigging you would use for the job. Is this the only rigging which could be used advantageously?

20. The men represented in Figure 5 are each exerting a force of 50 pounds. Neglect friction and find:
   (a) The force which the snub must stand.
   (b) The tension in each rope.
   (c) The force acting on the gun mount.
   (d) The force which the wall is called upon to withstand.

Assume the resistance found in (c) and, taking account of friction, find (e) the force which each man must exert.

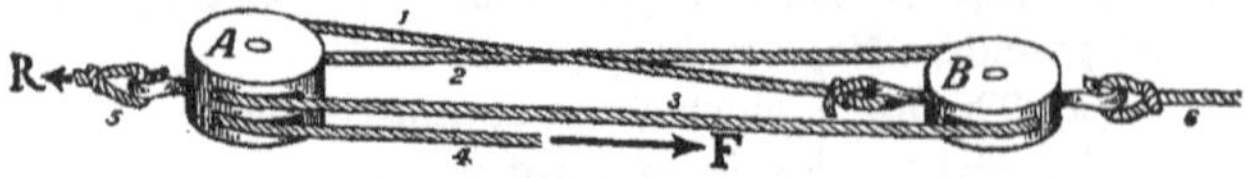

Luff Tackle

LUFF TACKLE

21. Show that two different mechanical advantages are possible with the Luff Tackle. What is their ratio?

22. Does the pulley P aid in securing the mechanical advantage attained in the rigging shown in Figure 6?

23. An unloading derrick is to be hastily constructed and the blocks and tackle for but one Luff Tackle are available. What mechanical advantage can be secured? Sketch the layout.

24. To what use would you put an additional single block?

25. What is the value of the mast-to-boom Luff Tackle in such a situation as is shown in Figure 6?

26. The drum operating the mast-to-boom tackle shown in Figure 6 is geared to have one fourth the speed of the other drum. What simpler mast-to-boom tackle would you use and why? Why is it desirable to use the Luff Tackle in case the gearing is the same?

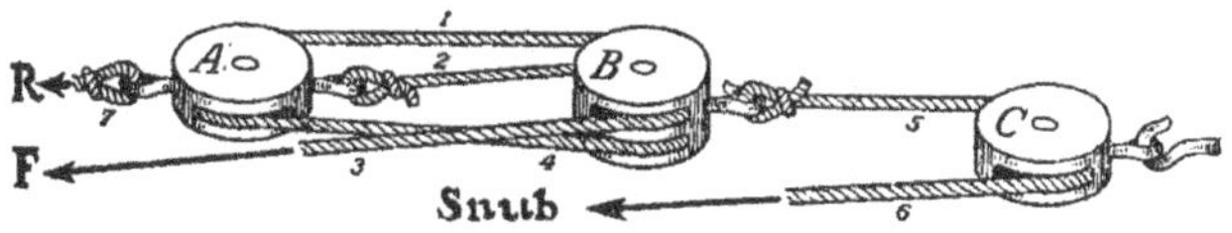

Port Tackle

Fig. 7

PORT TACKLE

27. What are the practical advantages of the Port Tackle?

28. (a) If the barrels in Figure 7 each weigh 30 pounds, what force must the man exert? Neglect friction.

(b) How strong must the rope be which supports the double block? Assume that the man is pulling at an angle of 30 degrees to the horizontal.

29. (a) Why must the man in Figure 7 stand in line with the mast?

(b) What is the practical difference whether he stands close to or far away from the mast?

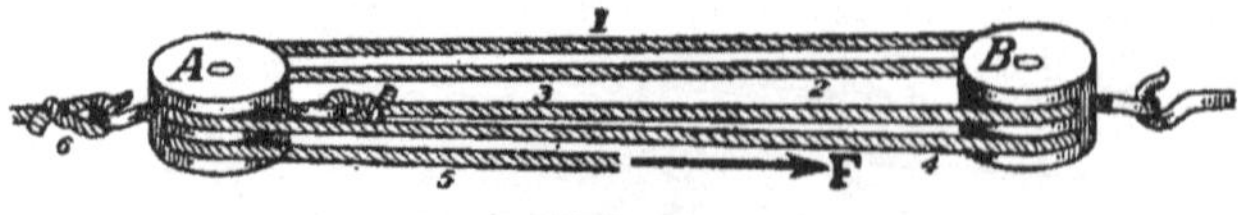

Double Luff (2 Fold Tackle)

Fig. 8

DOUBLE LUFF OR TWO FOLD TACKLE

30. Sketch two ways of rigging the Double Luff to drag a field piece up a hill. Compare the two as to relative mechanical advantage, and practical utility.

31. Show by sketch how you would use the Double Luff for hoisting.

32. What part of the total load is carried by each of the returns? What is the relation between the mechanical advantage and the number of returns supporting the load? What percentage of the effective mechanical advantage is lost as a result of friction?

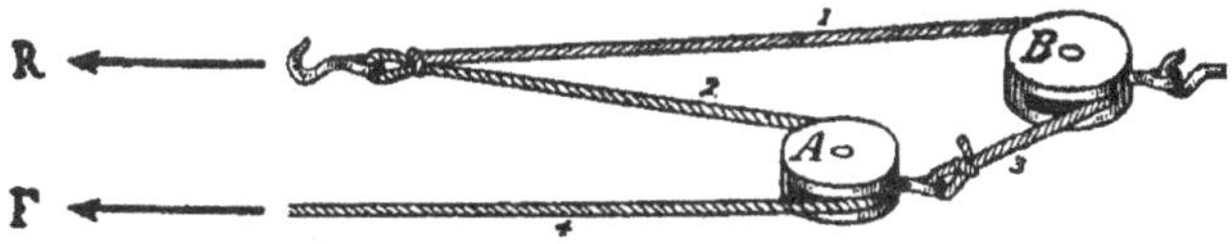

Single Burton (Type A)

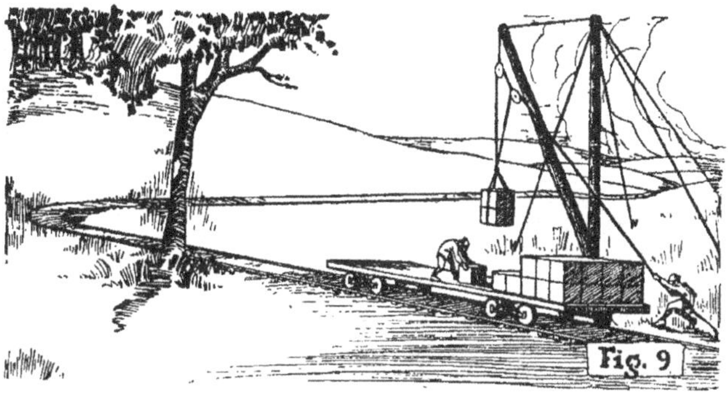

SINGLE BURTON, TYPE A

33. What is the mechanical advantage of the Single Burton, Type A?
34. Compare the mechanical advantages of the Whip on Whip and the Single Burton, Type A.
35. Why would the Single Burton, Type A, be of no practical utility in the job shown in Figure 5 if the men are able to move briskly when using the Whip on Whip?
36. What are the relative tensions in the two ropes of the Single Burton, Type A?
37. Show that the job of hauling the I beam to the tree (Figure 4) can be done more quickly with the Gun Tackle than the Single Burton, Type A.
38. Which is the faster, the Gun Tackle of Figure 3 or the Single Burton of Figure 9? With which tackle can one man lift a heavy load more easily? In which is the friction loss the greater?
49. Compare the different hoisting tackles which have been studied

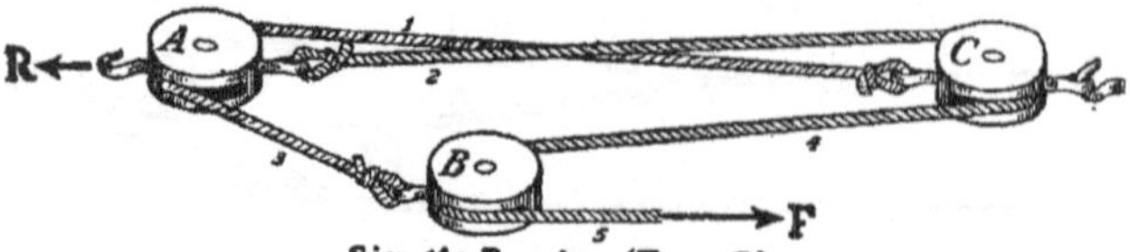

Single Burton (Type B)

as to: (a) height of lift, (b) strength of rope required, (c) length of rope required, (d) force required (man power), (e) convenience and speed.

SINGLE BURTON, TYPE B.

39. What is the mechanical advantage of the Single Burton, Type B?
40. What are the relative tensions in the two lengths of rope?
41. If there were plenty of the stronger rope, how would you change the rigging of the Single Burton in order to increase the mechanical advantage? Is there any loss of speed through this change?
42. How much oftener would the new rigging have to be overhauled?
43. Show that a still greater mechanical advantage could be secured with the equipment of Figure 10, if one were permitted to cut one of the ropes.
44. What is the relation between the mechanical advantage of the Burton, and the number of returns supporting the load?

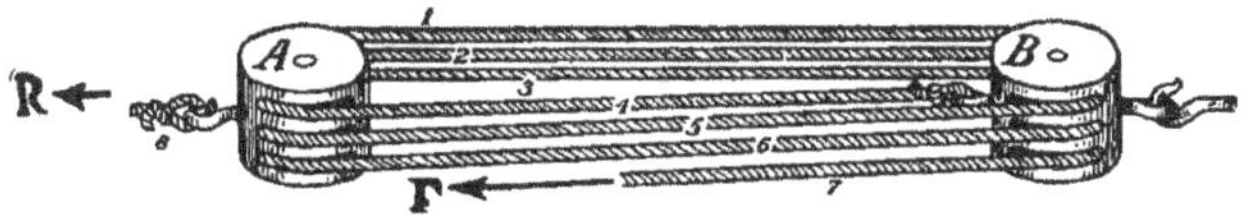

Three Fold Purchase

THREE FOLD PURCHASE

45. Why is the Three Fold Purchase of greater practical utility for high lifts than the more mechanically advantageous Burton?

46. Why would the Luff on Luff (Figure 17) be unsuitable for such hoisting work?

47. How strong a cable would be necessary to raise a one ton girder with the Three Fold Purchase?

48. How heavy a load could two men conveniently raise with the Three Fold Purchase?

FourFold Purchase

FOUR FOLD PURCHASE

50. In the Three Fold Purchase what is the percentage loss in effective mechanical advantage caused by friction?
51. How does the friction loss in the Three Fold Purchase compare with the friction loss in a Single Burton with the same mechanical advantage?
52. How much of a load can be lifted by the arrangement shown in Figure 12? The drum diameter is 8 inches, the length of the crank handles 16 inches.
53. Compare the Four Fold Purchase and the Three Fold Purchase for mechanical advantage and practical utility.
54. How much more mechanical advantage would the Four Fold Purchase have than the Two Fold Purchase if friction could be neglected? Taking account of friction what is the difference in the mechanical advantages?
55. How strong must be the fastening of the single block, the fastening of the upper quadruple block, the lower quadruple block, in Figure 12?

56. In the Three and Four Fold Purchase what is the relation between the number of returns supporting the load and the mechanical advantages? What part of the total load is carried by each of the returns?
57. How strong must the cable of a Four Fold Purchase be in order to just support a load of two tons? Take account of friction and find the strength necessary if the load is to be raised.

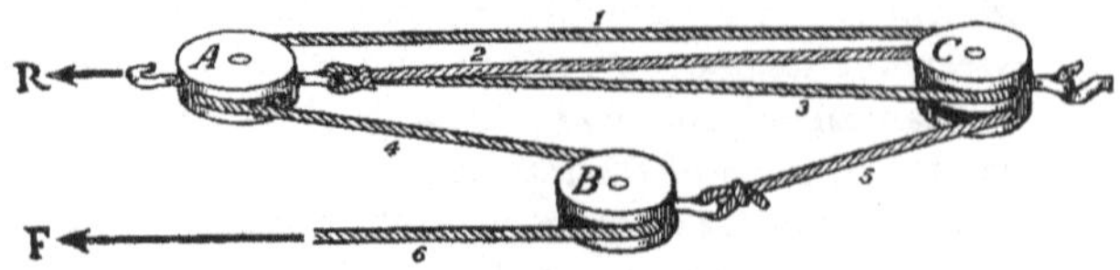

Double Burton (Type A)

Fig.13

DOUBLE BURTON, TYPES A AND B

58. What is the difference in the effective mechanical advantage of a Four Fold Purchase as found by taking account of friction by each of the two methods mentioned on page 33.
59. What are the respective mechanical advantages of the Double Burton, Types A and B?
60. How is the load distributed among the four returns in the Double Burtons, Types A and B?
61. Explain how the difference in mechanical advantage in Types A and B of the Double Burton is secured.
62. Is there any relation between the number of returns and the mechanical advantage?
63. Show that the Burton allows you to make use of a short length of rope to advantage.
64. Compare the Double Burtons, Types A and B, as to: (a) mechanical advantage, (b) distance load can be moved without overhauling, (c) speed.

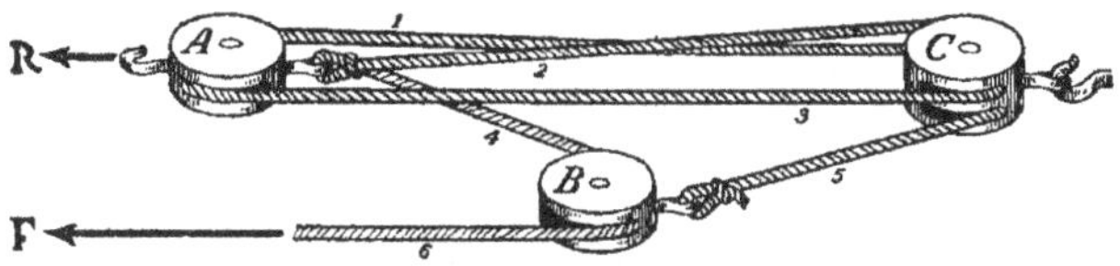

Double Burton (Type B)

Fig. 14

65. Compare the two Double Burtons, Types A and B as to percentage loss of effective mechanical advantage, as a result of friction.
66. Change the Double Burton, Type B, by attaching the standing end of the first rope to a snub instead of to the tackle block. Give your judgment as to comparative practical utility considering: (a) mechanical advantage, (b) distance load can be moved without overhauling, (c) speed.
67. Using the equipment of Figure 13 and assuming that you have a sufficient length of rope rig the blocks into a two-fold purchase. Compare with the Burton as to: (a) mechanical advantage, (b) distance load can be moved without overhauling, (c) speed.
68. Compare the losses in effective mechanical advantage of a Double Burton, Type A, and a two-block tackle with the same theoretical mechanical advantage. Which shows the greater loss? By what percentage?

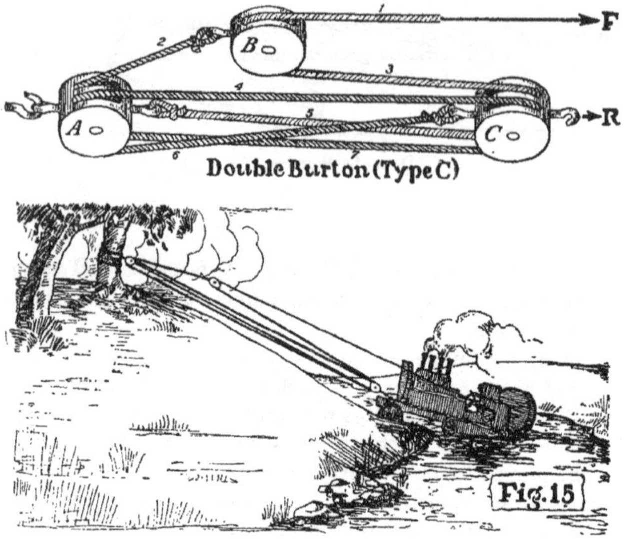

DOUBLE BURTON, TYPE C

69. What is the mechanical advantage of the Double Burton as shown at the top of Figure 15? As shown in actual use in Figure 15?

70. (a) Find the loss of effective mechanical advantage caused by friction in the rigging shown at the top of Fig. 15.

(b) Compare with the loss in a two-block tackle with the same theoretical mechanical advantage.

71. Consider the practical situation shown in Figure 15. Why would consideration of practical utility make you dispense with the lighter rope as soon as possible and apply the heavier rope to the drum as soon as possible? How much mechanical advantage would be lost?

72. Consider the practical situation shown in Figure 15.

(a) If you had plenty of heavy rope how would you rig for this job and why?

(b) What is the mechanical advantage as shown in the figure?

(c) What would be the mechanical advantage if you proposed rigging?

(d) Why is the mechanical advantage not the most important consideration in this situation?

73. A heavy load is to be moved with the Double Burton, using one single and two double-blocks.

(a) Sketch the rigging that will give maximum distance without overhauling.

(b) Put this same equipment into a rigging that will give a mechanical advantage of six and require no overhauling.

(c) Rig as a Double Burton to enable the load to be moved with the fewest number of men.

(d) Assuming that three suitable lengths of rope are available rig for the greatest possible mechanical advantage.

(e) Compare the riggings for practical utility.

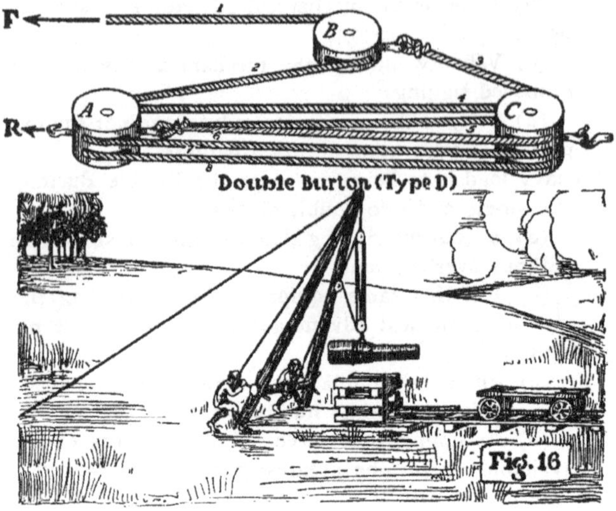

DOUBLE BURTON, TYPE D.

74. What is the mechanical advantage of the Double Burton shown in Figure 16?

75. If the blocks are separated ten feet, how high can the cannon be raised before overhauling will be necessary?

76. If there were sufficient heavy rope to rig for maximum mechanical advantage, how much easier would it be for the men? Assume that they each are exerting a force of 60 pounds.

77. Rig the equipment of Figure 16 as a Three Fold Purchase. Compare with other possible riggings as to practical utility.

78. What is the effective mechanical advantage of the Double Burton, shown in Figure 16?

79. The cannon in Figure 16 weighs 1,800 pounds; the lighter rope can stand only 200 pounds. Sketch the rigging that will require the least overhauling. What is its mechanical advantage? How strong must the heavier rope be?

80. What will be the stress on the rope supporting the upper block? On the hook of the lower block?
81. In your opinion how heavy a cannon could be raised using the rigging shown in Figure 16? Study the sketch and make necessary estimates and assumptions.
82. What is the maximum mechanical advantage you can secure with the equipment of Figure 16 if you have another length of rope?
83. If you have a load of 1,200 pounds to move a considerable distance and have plenty of rope that will stand 200 pounds, how will you rig using the blocks shown in Figure 16?
84. Why is a Three or Four Fold Purchase preferable to the more mechanically advantageous Burtons for high lifts? In what respects are the Burtons more advantageous when the lifts are small?
85. Is the loss of effective mechanical advantage due to friction greater in the case of the Burtons or two-block tackles?

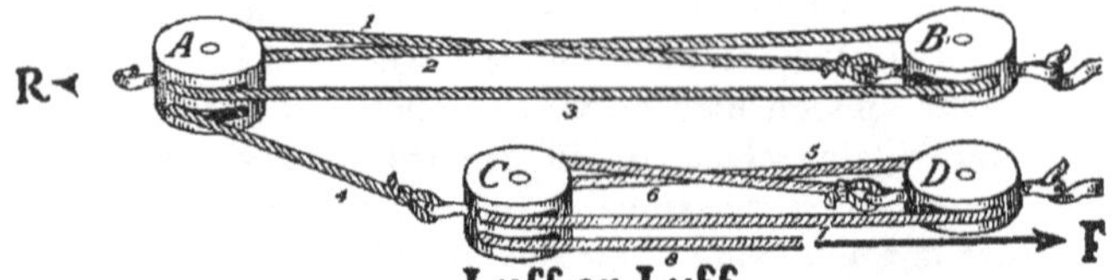

Luff on Luff

Fig. 17

LUFF ON LUFF

86. What is the mechanical advantage of the Luff on Luff? Why would it usually not be worth while to rig the equipment shown in Figure 17, so as to secure a greater mechanical advantage?

87. How much of a pull can be exerted on the gun carriage shown in Figure 17? What is the strain on the first rope? On the second rope? What is the strain on each block?

88. What is the great practical utility of the Luff on Luff? Explain why a piece of very heavy rope or cable is necessary in order to secure this advantage.

MISCELLANEOUS PROBLEMS ON BLOCKS AND TACKLE

89. A carload of 100 pound barrels is to be unloaded to a platform twenty feet above the tracks. You are assigned to the job with three helpers, two of whom are to do the pulling. What tackle will you use? State your reasons for the choice and sketch the rigging.

90. If you had only one helper for the job what change would you make in the tackle rigging?

91. A carload of 30-pound boxes is to be unloaded to a platform twenty feet above the tracks. You are assigned to the job and given one helper. What tackle will you use? State your reasons for the choice.

92. Your equipment consists of a double and a single block and plenty of rope capable of standing a force of 200 pounds. Rig the tackle so as to secure the greatest mechanical advantage and state the load which could be lifted.

93. Your equipment consists of a double and two triple blocks and plenty of rope capable of standing a force of 200 pounds. Rig the tackle for the greatest mechanical advantage and state the load which could be lifted.

94. Suppose you had plenty of men available and plenty of tackle, what rigging will you use to drag a cannon 50 feet, assuming that a force of 1,000 pounds is necessary and the rope can stand 400 pounds?

95. You must drag a heavy steel girder over the ground with the fewest men practicable. The available equipment consists of two double blocks and one single block and two ropes, one of 1,000 pounds tensile strength, the other of 300 pounds strength. Sketch the rigging you would use, and state the maximum safe load of the proposed arrangement.

96. What equipment and source of power would you prefer to employ to raise a one-ton safe to a third-story window?

97. What equipment and rigging would you prefer to employ to raise heavy marble cornice blocks to position?

98. A cannon is to be moved to a mounting 100 feet away. The equipment consists of 110 feet of rope which will stand a force of 200 pounds and 110 feet of rope which will stand 400 pounds, together with two single blocks.

(a) Show that if the resistance offered by the cannon is 600 pounds it can be moved half the distance without overhauling.

(b) Show that it can be moved practically the whole distance without overhauling if the resistance is 400 pounds.

(c) How many men will be required to each case?

(d) If you had plenty of the larger rope how would you rig the tackle for the fewest men? How many men would be needed?

(e) Could you move a cannon offering a resistance of 800 pounds with this equipment? If so, sketch the rigging. How many men would be required?

99. A cannon is to be dragged 100 feet to the base of a wall in which there is a tackle bolt. A force of 1,000 pounds is required. If you had one horse with which to do the job and your choice of blocks and tackle, what equipment would you employ? Specify the blocks, and lengths and strength of ropes.

100. What rigging would you select for the job shown in Figure 5 if you had to carry the equipment some distance? The force required is 150 pounds, the haul is 75 feet. Specify blocks, and lengths and strengths of ropes.

101. What rigging would you select for the job shown in Figure 5 if you had a horse for the job? The force required is 1,000 pounds, the haul is 75 feet. Specify blocks and lengths and strengths of ropes.

102. What rigging would you select for the job shown in Figure 5 if you had to do it with four helpers? The equipment must be carried a quarter of a mile. A force of 1,200 pounds is required. Specify blocks, and lengths and strengths of ropes.

103. What rigging would you use for the above job if you had a horse?

104. A log is 150 feet from a wall against which it is to be placed. Two pieces of rope each 160 feet long are available and the necessary blocks. How will you rig to get the greatest speed? How will you rig so as to use the fewest number of men?

105. Devise the cheapest practical hoist for a two-story building, using an extension of the ridge pole for the support, which will enable one man to handle 200-pound barrels.

106. Draw a suitable rigging for pulling a 25-foot launch out of the water for the winter.

107. What is the usual arrangement of the rigging of a pile driver with a 2,000 pound hammer? What considerations are involved in the choice of such a rigging?

108. With a winch capable of exerting a force of 1,000 pounds on the cable, how heavy a load could be lifted using the Three Fold Purchase? Allow for friction.

109. How fast could a 10-H. P. motor raise the load, assuming a force at the winch of 1,000 pounds?

110. Show that a force of more than 1,000 pounds is necessary while the load is gaining speed. What will be the average force if it acquires speed in the first half second?

111. An "A" frame is used to raise building material. There are two cranks, radius 16 inches, the gear ratio is 1 to 10, the diameter of the winding drum in 8 inches. What type of rigging would you employ if two men could be conveniently assigned to the work? If the men are to raise a 1-ton block of marble, what rigging should be employed. How long a rope would be required for a four-story building? How strong a rope?

112. Draw a diagram of the guy ropes for the "A" frame assuming available ropes have tensile strength of 500 pounds.

113. Draw a diagram of the rigging used in house moving, giving the reason for the choice of the different parts. How is the rigging accommodated to houses of different weights?

114. You must raise a 25-ton steam turbine from a car and lower it 10 feet to its foundation. You have available any length of cable of 20,000 pounds tensile strength and such blocks as you need. Draw a sketch of the rigging. What applied force will be necessary?

115. You must raise a 25-ton steam turbine from a car and lower it 10 feet to its foundation. You have available 25 feet of cable with 20,000 pounds tensile strength, two heavy single blocks and two light double blocks. You must buy sufficient cable of 5,000 pounds tensile strength. Show that with a little over 80 feet you can do the job without overhauling. Sketch the rig.

116. A launch is to be pulled out of the water. The necessary force is one ton. There is a large double block and two large single blocks. There is sufficient rope which will probably stand 500 pounds, but should be strained as little as possible. There are two small double pulleys and several single pulleys and plenty of rope that will stand 100 pounds. Rig so that one man can do the job of pulling.

117. You must raise a 25-ton steam turbine from a car and lower it 10 feet to its foundation. You have available 19 feet of cable with 20,000 pounds tensile strength, one heavy block, plenty of cable of 5,000 pounds tensile strength and several light blocks. Show how the job can be done without overhauling.

118. A howitzer weighing 10 tons is to be lifted from a road to a flat car. The road is so situated with respect to the car that there will be a lift of 5 feet and a swing of 10 feet in order to place the gun on the car. Arrange a single boom derrick and blocks and tackle to do the job. Specify strengths of tackle, ropes and guy ropes. With your rigging how much force will be required? How will you supply it?

PART III.
AERIAL BOMBING

AERIAL BOMBING

GLOSSARY OF TERMS

AIR-LAG—The horizontal distance, at any height, which the bomb lags behind a vertical line dropped from the airplane, assuming the airplane to continue at uniform velocity after dropping the bomb.

AIR-SPEED—The speed of the airplane relative to the air. It is read on an air-speed indicator.

BACK-SIGHT—A sighting rod or wire on a bomb-sight which has to appear in line with the fore-sight and the target at the correct moment for releasing the bomb.

BOMBING ANGLE—The angle between the vertical and the line joining the airplane and target at the correct moment for release of the bomb.

BOMB-SIGHT—An instrument designed to indicate the correct instant at which a bomb should be released in order to hit the target.

DRIFT-ANGLE—The angle between the fuselage and the direction of motion of the airplane relative to the ground. The angle between the direction of the air-speed and the direction of the ground-speed.

DRIFT-BAR—A bar which can be set to indicate the direction of motion of the machine over the ground.

FORE-SIGHT—The transverse sighting rod or wire on a bomb-sight furthest from the observer's eye.

GROUND-SPEED—The speed of the airplane relative to the ground. It is determined by the air-speed and wind-speed.

TIME-LAG—The difference between the actual time of fall of a bomb and the time of fall from the same height in vacuum.

TRAIL—The horizontal distance from the spot where the bomb strikes the ground to the point it would have reached in the same time under vacuum conditions.

TRAIL-ANGLE—The small angle by which the bomb, as viewed from a uniformly moving aircraft, appears to trail behind the vertical. Specifically, the angle subtended at the airplane by the trail.

TRAJECTORY—The path of the falling bomb.

WIND-SPEED—The speed of the wind relative to the ground.

AERIAL BOMBING

Aerial bombing operations fall under two heads, those in which it is not necessary to hit any particular object, and those in which it is essential to destroy some definite target. Examples of the firstclass are the German raids on London, the reprisal raids and the nightly low altitude raids near the front. The avowed aim of this type of raid is to destroy the morale of the enemy rather than to damage any object of direct military importance. The second type of bomb raid is the expedition which is considered to have failed if some special target is not destroyed. Examples are, raids to destroy railway junctions or bridges, munition factories, lock gates, and the like. These raids were seldom completely successful as it was not possible to place more than a small percentage of bombs near the objective.

Realizing the tremendous advantage which would be gained by the army which could first secure accuracy in bomb fire the various countries engaged in the war were and are still active in the development of bombs and bomb-sights. The problem divides itself naturally into two parts, bomb-sight design and bomb design, the latter including the study of bomb trajectories.

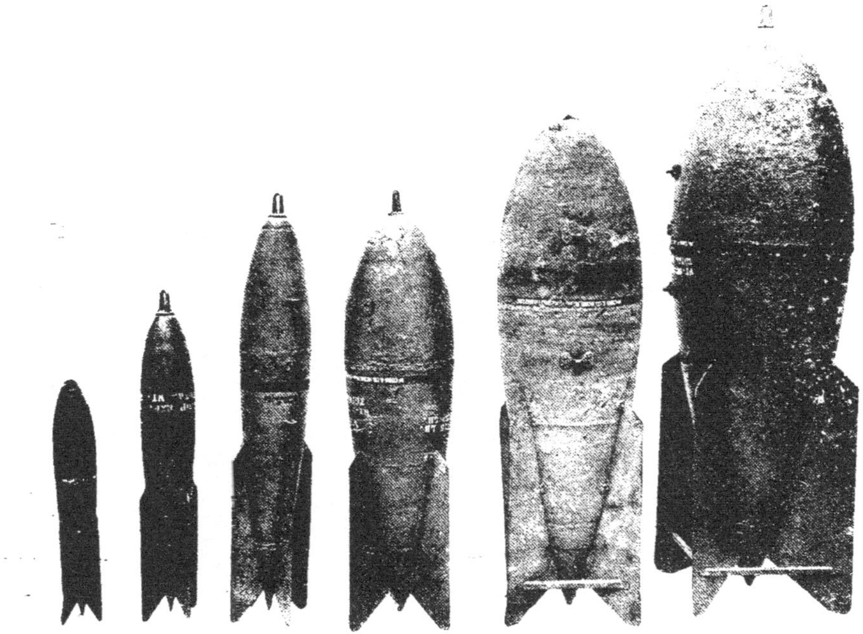

Figure 1. Group of Demolition Drop Bombs.
Weight 25, 50, 100, 250, 1000 pounds; length from 2½ to 5½ feet.

BOMB TRAJECTORIES

At Langley Field, Virginia, the Air Service has conducted, under the direction of Dr. A. W. Duff and Capt. L. P. Sieg, extensive experiments upon the trajectories of drop-bombs. This work is peculiarly interesting to students of physics as it is the first experimental data, obtained by photographic means, on the motion of a falling body from a height of more than a few feet.

The bombs are fitted with special electric lights of about 2,000 candle power, carried at the tail between the guide vanes. They are dropped at night and appear like brilliant shooting stars. The path of the light is photographed upon the plates of two cameras at the ends of a base line 2630 feet long. The cameras take the place of surveying instruments and may be regarded as equivalent to two transits.

In order that the true path of the bomb may be determined, it is necessary that corresponding points for the two plates be identified. Moreover, the time element must be introduced if velocities and accelerations are to be found. Both results are attained by producing breaks in the traces on the photographic plates by simultaneously closing the shutters of the two cameras at intervals known to an accuracy of two or three thousandths of a second.

The plates having been accurately measured, and a number of small but essential corrections applied, it is possible by suitable calculations to determine the actual path of the bomb in space, and its position at any instant, to an accuracy of about two feet. The velocity and acceleration of the bomb, at any point of the trajectory, can also be found to a high degree of accuracy. It is needless to explain that such information is of the utmost value in the study of air resistance.

The immediate information gained is the trail-angle and the time-lag of the bomb. From these can be computed all desired information concerning drop-bomb trajectories.

The airplane also carries an electric light which makes its trace upon the plate. In this way the path of the airplane and its ground-speed are determined. Similar records at different altitudes together with air-speed and compass readings make it possible to find the wind velocity at the different levels. Due to variations in the wind velocity and other variable influences such as lack of perfect symmetry in the bomb, the bomb always deviates a little from the plane in which it starts to fall and a three dimensioned diagram is required for the trajectory. This deviation has been ignored in Figure 2 in which is shown the trajectory of a Mark III 50-pound stream-line bomb dropped at a height of 5,539 feet from an airplane having a ground-speed of 98.28 feet per second.

PROBLEMS

1. Where would the bomb of Figure 2 have struck the ground if there had been no air resistance?
2. From the data on the trajectory of the Mark III 50-pound bomb, given in Figure 2, determine the trail.
3. Determine the air-lag at a point half way to the ground. Compare with the trail at the ground.
4. (a) Determine the air-lag at four or five points and for each point find the trail-angle.

   (b) Does the trail-angle increase or decrease as the bomb descends?

   (c) It is customary to assume that the trail-angle of a bomb is constant for different heights of release. Is this a reasonable assumption?
5. What would have been the time of fall of the bomb of Figure 2 if there had been no air resistance?
6. What is the time-lag, in seconds, of the Mark III 50-pound bomb whose trajectory is plotted in Figure 2?
7. What is the bombing angle for the actual condition revealed in Figure 2?
8. What would be the bombing angle for a bomb with twice the trail-angle of the bomb whose trajectory is given in Figure 2?
9. What would be the bombing angle for a bomb with twice the time-lag of the bomb whose trajectory is given in Figure 2?
10. What would be the bombing angle if the bomb were dropped in vacuo under the conditions represented in Figure 2?

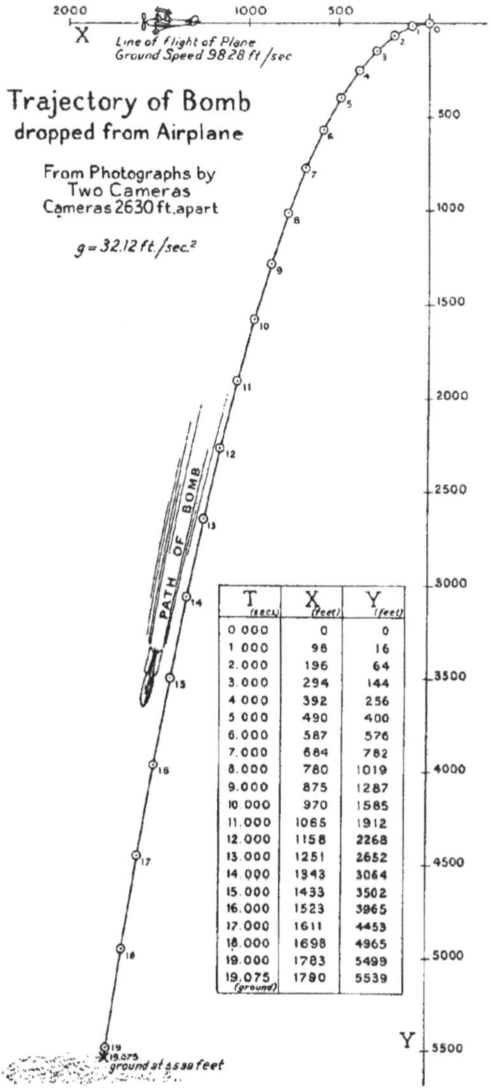

T (secs.)	X (feet)	Y (feet)
0.000	0	0
1.000	98	16
2.000	196	64
3.000	294	144
4.000	392	256
5.000	490	400
6.000	587	576
7.000	684	782
8.000	780	1019
9.000	875	1287
10.000	970	1585
11.000	1065	1912
12.000	1158	2268
13.000	1251	2652
14.000	1343	3064
15.000	1433	3502
16.000	1523	3965
17.000	1611	4453
18.000	1698	4965
19.000	1783	5499
19.075 (ground)	1790	5539

Figure 2

11. What would be the bombing angle if the Mark III 50-pound bomb were released from a plane with a ground-speed of 120 miles per hour at an altitude of 18,000 feet? At 6,000 feet?

12. What would be the bombing angle if the Mark III 50-pound bomb were released from a plane with a ground-speed of 30 miles per hour at an altitude of 18,000 feet? At 6,000 feet?

13. What would be the bombing angle if an airplane were flying at 12,000 feet with air-speed of 60 miles per hour in a head wind of 60 miles per hour?

14. Upon which does the trail-angle depend, the air-speed or the ground speed? Explain fully for; (a) flight against the wind, (b) flight with the wind, (c) flight across the wind?

15. The conditions are the same as those represented in Figure 2 except that the plane, due to "bumpy" air, is ascending at the rate of one foot per second at the instant of release in addition to the horizontal ground-speed of 98.28 feet per second. Consider only the effect of this upon the trajectory and proceed as outlined below to find how much the bomb will miss the target, i. e., how far from the point of impact indicated in Figure 2 it will hit.

(a) Give good reasons for assuming that the trajectory will be practically identical with that shown in Figure 2, after the path of the bomb has become parallel to the ground?

(b) What is the actual speed of the bomb at the instant of release?

(c) At what angle to the horizontal is it moving at the instant of release?

(d) Give good reasons for assuming that air resistance can be neglected in computing that part of the trajectory passed over before the path of the bomb becomes horizontal?

(e) How high above the point of release will it ascend?

(f) How long will it take it to reach this height?

(g) How far will it have traveled horizontally during this time?

(h) What angle with the horizontal is made by the path of the bomb at the instant it is at its greatest height?

(i) What will be its ground-speed at this instant?

(j) How high above the ground will it be?

(k) In what respects will the trajectory beyond this point differ from the trajectory shown in Figure 2?

(l) How far short of the ground will the bomb be when it has finished the part of its trajectory corresponding to the trajectory of Figure 2?

(m) In covering this vertical distance how far will it go horizontally?

(n) What is the total horizontal distance covered by the bomb?

(o) What was the horizontal distance planned upon when the bomb was released?

(p) By what distance does the bomb miss the target?

16. The conditions are the same as those represented in Figure 2 except that at the instant of release the airplane is descending at the rate of one foot per second in addition to its horizontal ground-speed of 98.28 feet per second. Consider only the effect upon the trajectory and find how much the bomb will miss the target, i. e., how far from the point indicated in Figure 2 it will hit.

17. An airplane is flying with an air-speed of 60 miles per hour in a head-wind of 30 miles per hour.

(a) What angle does the plane in which the bomb starts to fall make with the fuselage? With the direction of the ground-speed?

(b) If lack of symmetry in the bomb is neglected, will the path of the bomb remain in the vertical plane in which it starts to fall?

18. The airplane is flying with an air-speed of 60 miles per hour in a cross-wind of 30 miles per hour.

(a) What angle does the plane in which the bomb starts to drop make with the fuselage?

(b) What angle does this plane make with the direction of the ground-speed?

(c) Will the path of the bomb remain in the vertical plane in which it starts to fall or will the bomb be turned from its course? Consider for wind uniform all the way down and for variable winds. Neglect effects of lack of symmetry in the bomb.

19. Why is it that the trail depends upon air-speed rather than ground-speed? Why does the wind-speed have no influence on the trail?

20. A 50-pound bomb is released horizontally with a velocity of 98.28 feet per second from a height of 5,539 feet and is found to travel 1,790 feet horizontally before reaching the ground, the time of descent being 19.075 seconds.

(a) Calculate the mean vertical acceleration.

(b) Compare the mean vertical acceleration with that of gravity which is 32.12 feet per second per second at Langley Field.

(c) Calculate the mean vertical air resistance.

(d) Calculate the mean horizontal acceleration.

(e) Calculate the mean horizontal air resistance.

21. A bomb is released at a height of 6,000 feet from an airplane on a glide of 5 degrees. The (projected) ground-speed is 100 feet per second. Calculate the time of fall of the bomb and horizontal range, neglecting air resistance. Assuming the values of air resistance found in Problem 20, calculate the trail and time-lag.

22. Which of the bombs illustrated in Figure 1 has the smallest time-lag? Why?

23. Two bombs are released at the same instant, one hung vertically by its tail, the other suspended as shown in Figure 5.

(a) Which has the greater trail?

(b) Which swings the more easily into its trajectory?

(c) In your opinion which suspension is preferable?

24. In making dummy bombs for experimental purposes where must the center of gravity be located?

Figure 3. One of the Two Cameras at Langley Field, Va., Used to Photograph Bomb Trajectories.

THE LOOP

Problems dealing with the forces acting on an airplane while it is performing a loop are introduced at this point because the only experimental data ever obtained upon the actual performance of the airplane were secured by use of the same methods of photographic triangulation described in connection with bomb trajectories. Figure 4 shows the path of a Curtiss J N-4 H airplane executing a loop. The path is reduced to still air conditions.

Such information upon the actual performance of an airplane is of utmost importance in giving information concerning the stresses to which the airplane is subjected. Parts of an airplane which are amply strong enough to stand the stresses of plain flying may give way when subjected to unusually large forces produced by acrobatics. The forces which come into play when a plane suddenly changes its velocity, either in direction or magnitude, may become dangerously great. Some idea of the large factor of safety necessary in airplane construction can be gained from consideration of the following problems. In solving them it is to be assumed that the plane weighs 2,500 pounds and that the wings are the only supporting surfaces. Only the weight of the airplane and the force on the wings, in the direction of the radius of curvature, are to be considered.

PROBLEMS

25. Is it the air-speed or the ground-speed upon which depend the forces acting upon the wings of an airplane?
26. At what point does the airplane in Figure 4 have the maximum air-speed? The minimum air-speed?
27. At what point does the plane have the maximum ground-speed? The minimum ground-speed?
28. What is the vertical force acting on the wings when an airplane is flying on a level course?
29. What is the vertical force on the wings when an airplane is on an even glide of 20 degrees?
30. What is the vertical force on the wings when the plane is climbing steadily 400 feet per minute?

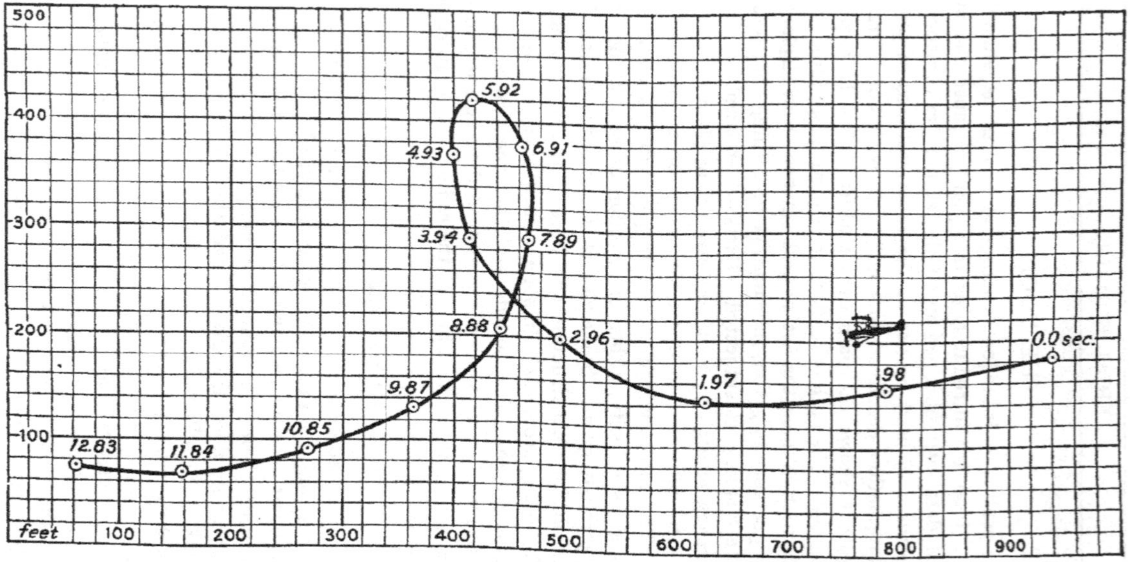

Figure 4. Path of J N-4 H Curtis Airplane making a loop in still air.
(Data taken from photographs made at night by two cameras 2630 feet apart.

31. If the elevator (horizontal rudder) is suddenly lowered so that the plane dives will the load on the wings be increased or decreased?

32. Why will the wings be called upon to withstand a greater load when the elevator is suddenly raised?

33. What force would be required to support the airplane at point 1.97 seconds, Figure 4, if it were stationary with respect to the earth? How much at 4.93 seconds? At 5.92 seconds?

34. What is the approximate radius of curvature of the path of the airplane at point 1.97 seconds?

35. What is the centrifugal force exerted by the airplane at position 1.97 seconds?

36. What is the amount of the total load upon the wings of the airplane at position 1.97 seconds? What is the direction of the force? How many times the weight W of the plane is the force?

37. What is the magnitude and direction of the force with which the air reacts upon the airplane at position 4.93 seconds? Express in terms of W, the weight of the plane.

38. What is the magnitude and direction of the force with which the air reacts upon the airplane at position 5.92 seconds? Express in terms of W, the weight of the plane.

39. What is the magnitude and direction of the force with which the air reacts on the airplane at point 9.87 seconds? Express in terms of W, the weight of the plane.

40. At what point in the path of the plane, as it makes the loop shown in Figure 4, are the wings called upon to withstand the greatest forces? How many times the weight of the plane does the force become?

41. At what point in the path of the plane, as it makes the loop shown in Figure 4, are the wings called upon to withstand the smallest forces? How many times the weight of the plane is the minimum force?

42. Sketch the path of the airplane making a loop in a head-wind of 30 miles per hour. In a head-wind of 60 miles per hour.

43. Sketch the path of the airplane making a loop down-wind in a wind of 30 miles per hour. Sixty miles per hour.

44. Does the shape of the path relative to the ground have any effect upon the reactions between air and plane? Explain.

THE EFFECT OF WIND ON THE MOTION OF THE AIRPLANE

The actual velocity of an airplane relative to the ground is the resultant of the velocity of the plane with reference to the air and the velocity of the wind. Needless to say, bombing operations necessitate accurate knowledge of both air-speed and ground-speed. The following problems are designed to bring out some of the more general considerations.

PROBLEMS

45. An airplane is flying with an air-speed of 70 miles per hour. The wind is blowing from the North with a velocity of 30 miles per hour. What is the ground-speed and direction of the ground-speed when the airplane is headed in the following directions: (a) North, (b) South, (c) East, (d) Northeast?

46. An airplane is flying with an air-speed of 60 miles per hour. The wind is blowing from the East with a velocity of 60 miles per hour. What is the ground-speed and direction of the ground-speed when the plane is flying in the following directions: (a) North, (b) South, (c) East, (d) West?

47. An airplane is flying with an air-speed of 120 miles per hour. The wind is blowing from the East with a velocity of 60 miles per hour. What is the ground-speed and the direction of the ground-speed when the plane is flying in the following directions: (a) North, (b) South, (c) East, (d) West?

48. An airplane flying at an air-speed of 70 miles per hour, at 335 degrees from North through East, is found by photographic triangulation to have a ground-speed of 55 miles per hour due North. Find the wind-speed, the direction of the wind, and the angle of drift of the airplane.

49. Same as Problem 48 except that the ground-speed is 30 miles per hour due East.

50. Same as Problem 48 except that the ground-speed is 50 miles an hour Southwest.

51. If the wind is blowing at 30 miles per hour and the air-speed of the airplane is 70 miles per hour what is the radius of the circle which can be drawn through the points which can be reached by the plane in one hour? Where is the center of the circle located with reference to the starting point?

52. An airplane with an air-speed of 60 miles per hour is to fly at an altitude where the wind is blowing from the East at 60 miles per hour.

(a) Draw the diagram which will show the ground-speed which it can make in any direction.

(b) Use the diagram to find in what direction the plane must be headed in order to actually travel Northwest. What ground-speed can it make?

(c) What ground-speed can the plane make North? West? East?

53. Make a similar diagram for a plane with an air-speed of 120 miles per hour in an East wind of 60 miles per hour.

(a) In what direction must the plane head to go Northwest? What ground-speed can it attain?

(b) What ground-speed can the plane make North? West? East?

BOMB-SIGHTS

On page 88 will be found a detailed description of the Wimperis-type bomb-sight in the form adopted by the U. S. Army. Certain points in connection with bombing can, however, be brought out more effectively if considered apart from the bomb-sight.

The problem of dropping a bomb at the right instant and right place resolves itself into two parts; getting the airplane on the right course and dropping the bomb at the right time. If the plane is not flying in the correct course the bomb will fall to the right or left of the target producing what is known as "line" or direction error. If the bomb is not dropped at the right instant the bomb will either fall short of the target or go beyond it. This is called "range" error.

The two sighting operations are entirely independent. The plane must first be brought into the right course. This having been accomplished the entire attention of the bomber is given to determining the correct instant for releasing the bomb.

THE DRIFT-BAR

A drift-bar is a device by which the pilot can find the angle between the direction of the air-speed and the ground-speed of the airplane. This angle, with the wind direction and air-speed, determines the wind-speed. A knowledge of the latter is necessary for a correct setting of the back-sights.

A drift-bar is nothing more than a horizontal bar capable of rotation about a vertical axis. See Figure 7, page 80.

PROBLEMS

54. A pilot is flying a straight course in still air. The drift-bar is set at zero, i. è., lies parallel to the axis of the fuselage. Do objects on the ground appear to move along the drift-bar from end to end or do they move across the bar at a definite angle?

55. What relation exists between the direction in which the plane is moving and the direction of the apparent motion of objects on the ground?

56. A pilot is flying a straight course directly into a strong head-wind. How do objects on the ground appear to move relative to the drift-bar, which is set at zero? The pilot turns the plane to the left. How does this affect the apparent motion of objects relative to the drift-bar?

57. The pilot adjusts the drift-bar until objects on the ground appear to move along the bar. What is the relation between the angle at which the drift-bar is set and the drift angle of the airplane?

58. An airplane is headed 115 degrees from North through East at an air-speed of 70 miles per hour in a 30 mile per hour wind from the Northeast. What is the angle of drift?

59. At what angle to the axis of the fuselage will the drift-bar be set when the plane is headed 20 degrees away from the direction from which the wind is blowing? When the plane is headed 20 degrees away from the direction toward which the wind is blowing?

60. The air-speed meter indicates 70 miles per hour, the compass shows that the plane is headed 90 degrees from North, through East, the drift-bar of the bomb-sight makes an angle of 20 degrees with the fuselage. What is the ground-speed and its direction?

61. A pilot is flying a straight course directly away from the wind. Which way must he turn in order to make use of the drift-bar of the bomb-sight shown in Figure 7?

62. A pilot flies with the wind in such a direction that objects seem to move along the drift-bar, set at zero, from end to end. The compass reads due North. The pilot heads the plane due East and adjusts the drift-bar until objects on the ground appear to move along it. If the drift-bar is found to be set at an angle of 30 degrees with its zero position and the air-speed has been constant at 70 miles per hour, what is the wind-speed and the direction of the wind?[1]

Figure 5. Drop Bombs in Place in Release Mechanism.

Figure 6. Types of American Bombs.
(From left to right, two fragmentation bombs, dummy bomb, two incendiary bombs, airplane flare. Note the arming devices like wind mills, which unscrew in flight and arm the bomb.)

63. A pilot flies into the wind in such a direction that objects on the ground appear to move parallel to the drift-bar, set at zero. The compass reads due North. The pilot heads the plane due West and adjusts the drift-bar until objects appear to move parallel to it. If the drift-bar is found to be set at an angle of 30 degrees with its zero position and the air-speed has been constant at 70 miles per hour, what is the wind-speed and the direction of the wind?[1]

[1] In the Wimperis type bomb-sight the wind-speed is found in the manner described. By a clever mechanical arrangement the setting of the drift-bar automatically separates the "up" and "down" back-sights the correct amount.

DIRECTION SIGHTING

A pilot using the Wimperis-type bomb-sight must approach the target either directly up-wind or directly down-wind. If at the instant the bomb is released the course of the plane is such as would carry it directly over the target there will be no "line" or direction error. Direction sighting is accomplished by means of the direction wire and the bubble of the lateral spirit level. The direction wire is fixed in position parallel to the axis of the fuselage. The lateral spirit-level is for the purpose of making the line of sight past the direction wire vertical. See Figure 7, page 80.

PROBLEMS

64. Why is it not necessary for the pilot's eye to be directly above the drift-bar when he is making a setting of the drift-bar?

65. What relation would the path of the airplane bear to the points on the ground which appear to move along the drift-bar, if the pilot's eye were in the vertical plane determined by the drift bar?

66. A pilot is flying a straight course in still air. How do objects on the ground appear to move relative to the direction wire?

67. If a wind is blowing, how must the pilot fly in order for objects on the ground to appear to move along the direction wire?

68. A pilot using the Wimperis-type bomb-sight must approach the target, at the moment the bomb is released, in a straight course, either up- or down-wind.

    (a) How will objects appear to move relative to the direction wire?

    (b) Where must the pilot's eye be placed, with reference to the direction wire, if the target is to appear to move along the direction wire when the approach is correct?

    (c) Where must the center of curvature of the lateral spirit level of the bomb-sight be located if a line of sight past the bubble and the direction wire is to be in a vertical plane?

69. What relation does the path of the airplane bear to the target if the target appears to move along the direction wire when the pilot sights past the bubble of the lateral spirit level? What significance has this in connection with the accuracy of the fire?

RANGE SIGHTING

A bomb-sight must indicate to the bomber the instant at which to release the bomb. This should be when the horizontal distance from the target is equal to the distance which the bomb will, after release, travel forward in its passage to the earth. It might at first seem that the bomb would drop vertically from the air-plane to the ground but such is not the case for the conditions under which a plane could be stationary with respect to the earth rarely occur. Normally a plane is moving over the ground at a speed of many miles per hour for it must keep in rapid motion with respect to the air or it will fall.

Let us first consider the simplest conditions, namely, still air and an airplane flying a straight level course at constant speed. At the instant the bomb is released it partakes of the full speed of the plane and if it were not for air resistance would travel forward, in its flight to the ground, a distance equal to that covered by the plane during the same time. Viewed from the ground its trajectory would be a parabola. Viewed from the airplane the bomb would appear to drop vertically to the earth and to burst immediately below the plane. Under these ideal conditions the horizontal distance between the target and the airplane at the instant at which the bomb should be released would depend solely upon the ground-speed of the airplane and its altitude, the latter, of course, determining the time.

As a matter of fact a great many other considerations are involved. If there is a wind, its speed and direction affect the speed of the airplane relative to the ground, and the bomb-sight must be provided with corresponding adjustments. In addition the presence of the air causes a departure from the theoretical parabolic trajectory in vacuo.

When viewed from the airplane the bomb appears to be moving backward at a slow and fairly uniform velocity. The total distance on the ground from the point of impact to the point vertically below the airplane at the time of impact is known as the trail. It is, of course, assumed that the airplane has moved with constant

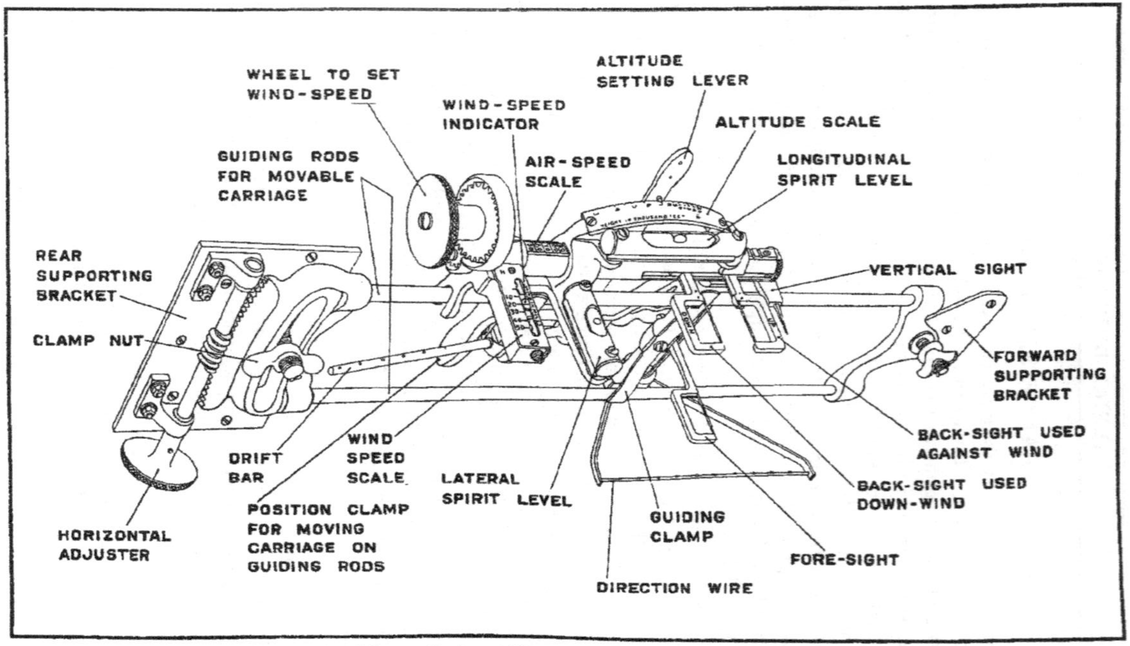

Figure 7. The U. S. Army Wimperis-Type Bomb-Sight Mark IA

velocity. This lagging of the bomb behind the vertical may also be measured by the trail-angle which is the angle at the airplane subtended by two lines joining it to the extremities of the trail. Another characteristic of the bomb is the time-lag or the difference in the time of fall along the actual trajectory and along the parabola of vacuum conditions.

PROBLEMS

Note:—In the Wimperis-type bomb-sight the adjustment of the back-sights for air-speed is made by sliding the two back-sights, and attached parts, forward or back a distance of .17 inch for each 10 miles per hour of air-speed. The fore-sight is movable up and down in a guide which is tilted at an angle of 2.8 degrees to the perpendicular, the lower end being nearer the tail of the machine. This corrects for the trail of the bomb, 2.8 degrees being a good average value of the trail-angle. The bomb-sight is leveled up in the fore and aft directions by means of the longitudinal spirit level.

70. Draw a diagram to scale showing the relative position of the fore-sight and the two back-sights, of the Wimperis-type bomb-sight, for bombing at an altitude of 12,000 feet. There is no wind and the airplane is flying with an air speed of 100 miles per hour. Assume a trail-angle of 2.8 degrees.
71. Draw a diagram, similar to that of Problem 70, to fit the following conditions. There is no wind and the airplane is flying at an air-speed of 50 miles per hour at an altitude of 3,000 feet. Assume a trail-angle of 2.8 degrees.
72. How much must the two back-sights be separated if the wind-speed is 10 miles per hour? Fifty miles per hour?
73. Draw a diagram, similar to that of Problem 70, to fit the following conditions. The airplane is flying at a height of 9,000 feet with an air-speed of 60 miles per hour in a 30-mile wind. In a 60-mile wind. In the latter case, which one of the back-sights is practically over the fore-sight?
74. Make a rough sketch showing the surface of the ground, the line of flight of an airplane, the actual trajectory of a bomb, and the theoretical parabolic trajectory in vacuo.

(a) Mark the position of the plane at time the bomb strikes the ground.

(b) Mark the point at which the bomb strikes the ground.

(c) Mark the point of theoretical impact, assuming the theoretical time of fall for the assumed altitude. What is the position of this point in relation to the airplane?

(d) Mark the point of theoretical impact, assuming that the bomb falls in vacuo for a time equal to the actual time of fall. What is the position of this point relative to the airplane? Why is this point below the surface of the ground?

(e) Mark the trail, trail-angle, and air-lag.

75. Draw or trace the trajectory of Figure 2 accurately and proceed as follows:

(a) Plot the position of plane at time of impact of the bomb.

(b) Show the point of theoretical impact, neglecting air resistance, computed on the basis of altitude.

(c) Plot the position of the point of actual impact, neglecting air resistance, computed on the basis of actual time of fall.

(d) Plot the theoretical path from a few computed points.

PROBLEMS ON ERRORS IN BOMBING

76. The accuracy of aerial bomb fire may be stated in terms of the circle within which 50 per cent of the bombs are likely to fall. With present sighting methods the radius of the circle is approximately 2 per cent of the altitude. What is the radius of the circle within which 50 per cent of the bombs will fall when dropped from a height of 6,000 feet? 12,000 feet?

77. The accuracy of aerial bomb fire is often expressed in terms of the angle at the airplane subtended by the line joining the target and point of impact. To what angle does the error mentioned in Problem 76 correspond?

78. A Wimperis-type bomb-sight is set at the ground for an elevation of 10,000 feet, an air-speed of 60 miles per hour and a wind-speed of 30 miles per hour, on the basis of information from the Meteorological Service. By how much will the bomb miss the target as a result of each of the following sources of error? Consider both up-wind and down-wind attack.

(a) The average error of altimeters (aneroid barometers) is about 300 feet at 10,000 feet.

(b) The maximum error of altimeters is about 1,000 feet at 10,000 feet.

(c) An air-speed meter may be in error as much as 5 miles per hour.

(d) The wind-speed as furnished by the Meteorological Service may be in error by 5 miles per hour.

(c) The direction of the wind may be wrong by as much as 10 degrees.

79. An airplane is flying at an altitude of 10,000 feet with an air-speed of 100 feet per second. The sight is taken correctly and the bomb released at the right instant but owing to "bumpy" air the plane is moving upward at the rate of 20 feet per second at the instant of release. What is the corresponding ground error?

80. An airplane with a ground-speed of 50 miles per hour drops a bomb when flying at an altitude of 10,000 feet. What is the maximum distance by which the bomb can miss the target as a result of the pitching of the plane through an angle of one degree? Consider only the effect due to inaccuracy of sighting due to the pitching.

81. If "bumpy" conditions cause a variation in the ground-speed of 0.6 miles per hour, what is the corresponding error for an airplane with a ground-speed of 60 miles per hour at an altitude of 10,000 feet?

82. With certain bomb-sights the ground-speed is determined by noting the time taken by an object to pass from an angle 15 degrees ahead of the plane to a point immediately below the plane. If the plane is flying at 10,000 feet altitude, at a ground-speed of 60 miles per hour, what is the maximum possible error in the determination of the ground-speed caused by a forward and aft pitching of the plane amounting to one degree?

83. The ground-speed is being found by measuring the time taken for an object to pass from an angle 15 degrees ahead to a position under the machine. If the ground-speed is found to be 50 miles per hour based on an altimeter reading of 10,000 feet which is 500 feet too low, what is the true ground-speed?

84. The bomb-sight is of the Wimperis-type. The pilot looks down past the bubble of the lateral spirit level and the direction wire and notices that he is going to pass 300 feet to the right of the target. He banks and turns to the left.

(a) Why is a line of sight past the bubble of the lateral spirit level and direction wire no longer vertical?

(b) If the pilot makes a perfect blank, what angle does the plane, determined by the wire and bubble, make with the perpendicular?

(c) If the pilot banks at as small an angle as 5 degrees, what "line" or direction error on the ground is indicated by the sight? Altitude of airplane 10,000 feet.

(d) Show that the pilot will probably, on coming out of the bank, find that he has turned too far.

85. Why should the lateral spirit level never be used in sighting unless the airplane is flying a straight course?

CROSS-WIND BOMBING

Attempts have been made to devise satisfactory sights for cross-wind bombing. Many interesting considerations arise which do not enter into the problem of up- and down-wind bombing with the Wimperis-type sight.

PROBLEMS

86. An important consideration that arises in cross-wind bombing is whether or not the airplane passes directly over the point at which the bomb strikes, assuming of course that the airplane continues in a straight course.

(a) How far from a vertical line through the point of impact does the plane pass if it is flying with an air-speed of 60 miles per hour, at a height of 12,000 feet, in a head-wind of 30 miles per hour?

(b) Same question and conditions except that the wind makes an angle of 30 degrees with the direction in which the plane is pointed.

(c) Same question and conditions as in (a) except that the plane is flying down-wind with the wind at an angle of 30 degrees with the fuselage.

87. The relation of the direction of the trail to the direction in which the airplane is headed at the moment the bomb is released is of great importance in cross-wind bombing. Find this relation by answering the following questions in regard to a plane headed Northwest and flying with an air-speed of 60 miles per hour in a South wind of 30 miles per hour.

(a) What is the direction in which the plane is actually moving? What is its ground-speed?

(b) At the instant the bomb is dropped what is its direction of motion?

(c) At the instant of release what is the velocity of the bomb in the direction of the wind?

(d) How does this compare with the wind velocity?

(e) What resistance will the bomb encounter in the direction of the wind?

(f) What will be the air-lag in the direction of the wind?

(g) From what direction will the air apparently strike the bomb? Is this direction the same as that in which the airplane is headed?

(h) What will be the relation between the direction of the trail and the direction in which the plane was headed at the moment of release?

88. Give evidence to show that the "trail" is always in the direction occupied by the fuselage at the instant the bomb is released. Consider the following case. A plane is headed Northwest and flying with an air-speed of 60 miles per hour in a South wind of 30 miles per hour.

(a) What is the direction in which the plane is actually moving? What is its ground-speed?

(b) At the instant the bomb is dropped what is its direction of motion?

(c) At the instant of release what is the velocity of the bomb in the direction of the wind?

(d) How does this compare with the wind velocity?

(e) What resistance will the bomb encounter in the direction of the wind?

(f) What will be the air-lag in the direction of the wind?

(g) From what direction will the air apparently strike the bomb? Is this direction the same as that in which the airplane is headed?

(h) What will be the relation between the direction of the trail and the direction in which the plane was headed at the moment of release?

DESCRIPTION OF THE U. S. ARMY WIMPERIS-TYPE BOMB-SIGHT MARK I A

A number of different types of bomb-sights have been developed and used. Experience has shown that those which embody a stop watch and require approach to the target in a straight course for a considerable distance are not used by pilots under fire. The Wimperis-type high-altitude bomb-sight was designed with the object of providing a sight which can be set almost instantly and permits the pilot to make the preliminary approach toward his target from any direction and in any manner he chooses, so long as he straightens out into a level straight course toward the target either directly up- or down-wind, for ten or fifteen seconds. It is a sight which pilots will use under fire.

The Wimperis-type bomb-sight in the form in which it was approved for production in the United States Army is shown in Figure 7, page 80. The sight is mounted on the right side of the fuselage in a position best understood by reference to the two spirit levels. This sight is used for both direction sighting and range sighting. Direction sighting is accomplished with the direction wire and the lateral bubble. The pilot swings into position, either up-wind or down-wind from the target, and straightens out into a level course directed toward the target. If he has chosen his position correctly a sight past the bubble of the lateral spirit level and the direction wire falls upon the target, which appears to move along the wire.

This is due to the fact that the center of curvature of the tube of the lateral spirit-level is in the line of the direction wire. Consequently the bubble is always directly above the line of the direction wire unless the plane is subjected to accelerations.

If the pilot finds that his course is not directed over the target he must turn quickly. This requires banking which makes the bubble useless until the plane straightens out again. It can then be determined whether the course was altered the right amount. (See Problems 84 and 85.)

Range sighting is accomplished by the use of three sight wires, the fore-sight and two back-sights, one of the back-sights for use when the airplane is flying against the wind and the other for down-wind attack.

The altitude adjustment is made by moving the fore-sight up and down by means of the altitude setting lever, the proper position of which is indicated on the altitude scale. The fore-sight also takes care of the adjustment for the trail-angle of the bomb. This is accomplished by giving the guide which carries the fore-sight a permanent tilt at an angle of 2.8 degrees with the vertical. This value is sufficiently close to the trail-angle of the bombs used with this sight.

The air-speed and wind-speed adjustments are made by moving the back-sights horizontally. The air-speed adjustment is made by grasping the wind-speed adjusting wheel and pushing all that part of the sight which has to do with the back-sights forward or back on the guides. Ten miles per hour air-speed is represented on the air-speed scale by a distance of .17 inch.

If the meteorological service is able to furnish information regarding the wind velocity at the point where the bombing is to be done the wind-speed adjustment is made before the airplane leaves the ground, by turning the wind-speed adjusting wheel until the drift-bar indicates the right figure on the wind-speed scale.

If the wind-speed is not known accurately enough from meteorological observation it can be determined by the operator. He first flies directly with the wind or against it, determining the proper direction by observing the drift of objects on the ground in relation to the direction wire. When objects seem to move along this wire the airplane must be flying in the direction of the wind.

The operator notes his course and turns at an angle of 90 degrees, by compass, so that the wind strikes the machine on the right, this being a right-hand sight. By operating the hand wheel he adjusts the drift-bar until objects on the ground appear to move along it. This operation automatically moves the "up" and "down" sight-wires apart and records the wind-speed on the wind-speed scale. The "up" sight is moved forward automatically until its distance from the zero point of the air-speed scale is proportional to the ground-speed for up-wind flight, the ground speed being equal to the wind-speed subtracted from the plane's air-speed. The "down" sight is moved back simultaneously and automatically until its distance from the zero of the air-speed scale is proportional to the ground-speed for down-wind flight, the ground-speed being equal to the wind-speed added to the plane's air-speed.

It is impossible to adequately describe the clever mechanical features of the Wimperis sight. It is hoped that every college desiring a Wimperis sight will be able to secure one from the Ordnance Department. (See note below.)

With the Wimperis-type bomb-sight the plane attacks either up- or down-wind and approaches the target on a straight level course which would carry it directly over the target. Just before the bomb is to be released the attention of the operator is directed entirely to the matter of range. At the instant the proper back sight, either "up" or "down" according as he is bombing up- or down-wind, comes in line with the fore-sight and the target the bomb is released. Great care must be exercised at this time to preserve the level of the plane and maintain uniform air-speed.

Note: Arrangements will probably be made so that U. S. Army Wimperis-Type Bomb-Sights Mark I A can be obtained from the Ordnance Department of the U. S. Army. Requests from institutions where there are Reserve Officers' Training Corps Units should be made through the Commandant, and addressed to the office of the Chief of Ordnance, United States Army, Washington, D. C.: Attention of the Training Section of the Administrative Division.

It is possible that moving picture films showing a falling bomb as seen from the airplane will also be available.

PART IV.

AERIAL PHOTOGRAPHY

AERIAL PHOTOGRAPHY

The great part which aerial photography played in the recent war is too well known to need to be elaborated upon. Its usefulness has by no means ended for it promises to be a very powerful aid to map making and will have many other applications as well.

The success which has been met in aerial photographic work has resulted from the painstaking study of a large number of physical problems. By no means all of these have been solved. For example, it still remains for some one to develop a satisfactory means of keeping a camera pointed vertically or, failing in that, of recording its angle with the vertical.

If a photographic map or mosaic such as is shown in preparation in Figure 2 is to be made, the airplane flies over the territory and vertical pictures are taken in rapid enough succession to overlap each other. The pictures so obtained are matched together, giving a photographic map of a strip of territory. If more territory is to be mapped than can be covered by a single strip, several strips are made by successive flights of the plane over parallel courses.

A great variety of types and sizes of camera for vertical use has been developed, ranging in focal length from 8 inches to 48 inches. Toward the end of the war, the British and American practice was standardized to correspond with the French, viz: 18 x 24 centimeter negatives, with 50 centimeter focal length lenses for use when a large territory has to be covered quickly, and a few 120 centimeter lenses for certain occasions when great detail is needed. These rather long focal lengths were adopted partly because of the great heights (frequently 18,000 to 20,000 feet) at which photographic reconnaissance had to be carried on over territory well protected by anti-aircraft batteries and enemy planes.

The development of the art of aerial photography has been largely dependent upon the application of principles of physics. The problems which are given below are based on some of the simpler aspects of the development work and upon the situations met in carrying out photographic work in the field.

GENERAL PROBLEMS

1. Exposures for a photographic mosaic of an area of 10 miles by 15 miles on a scale of 1 to 5000 are to be made using the American Type K-1 camera, Figure 7, which takes a picture 18 x 24 centimeters.

   (a) How many exposures are needed if 25 per cent overlap is required from picture to picture, from strip to strip, and beyond the edge of the area?

   (b) If the airplane flies at 70 miles per hour air-speed, what time interval must elapse between exposures if there is no wind?

   (c) Assuming a uniform rate of climb of 600 feet per minute and allowing 5 minutes for making a turn and starting back for a new strip, what flying time will be required? The camera is fitted with a 20-inch lens and is capable of 100 exposures in a flight. The flying field is 15 miles from the area to be photographed.

   (d) If there is a wind of 15 miles per hour and the airplane flies at an air-speed of 70 miles per hour, what will be the time interval between exposures for strips made flying against, and with the wind, respectively? The camera is placed so that the short side of the picture is parallel to the direction of the ground-speed.

   (e) What will be the time interval between exposures if the strips are made in a direction perpendicular to the wind by "crabbing" across the wind? The camera is turned so that the short side of the picture is parallel to the ground-speed.

2. How much more rapidly can the above territory be covered using a 20-inch focal length lens than with a 40-inch lens if the plane flies at the same altitude in each case? What will be the scale of the map made with the 40-inch lens?

3. How large an area can be photographed on a single plate from an airplane flying at 6,000 feet altitude, using an 18 x 24 centimeter plate and a lens of 50 centimeter focal length? Using a lens of 25 centimeter focal length?

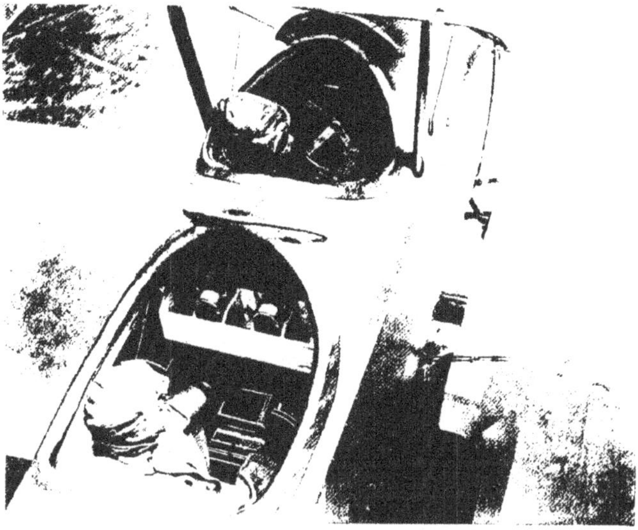

Figure 1. American Type L Camera in Operation.

Figure 2. Preparing a Photographic Map or Mosaic.

4. A photographic mosaic has been made with a 50 centimeter focal length lens with the plane flying at an altitude of 8,000 feet. More pictures are needed, to the same scale as the mosaic, at points of enemy activity. On account of the danger to a low-flying plane it is proposed to use a 120 centimeter lens. At what height must the pilot fly?
5. If photographic reconnaissance cannot be carried on below 18,000 feet, on account of anti-aircraft guns, what is the largest scale on which you can get a mosaic if you have lenses of 25, 50 and 120 centimeters focal length? Which lens is used?
6. If the operator knows that his camera is likely to deviate as much as 5 degrees from the perpendicular, what overlap must he plan for in order to be certain to obtain an overlap of 15 per cent in the finished prints; (a) using an 18 x 24 centimeter plate and a 120 centimeter lens, (b) using an 18 x 24 centimeter plate and a 50 centimeter lens?

TIME OF EXPOSURE

Exposures in aerial photography must always be very short. Not only does the ground-speed of the airplane put a definite limit on the allowable time of exposure, but the angular motion of the whole plane, i. e., rolling and pitching, and of the camera alone, caused by vibration, are sources of very great difficulty.

Lieutenant J. P. Brinsmade, of the Air Service, carried on extensive experiments upon the effect of vibration of the camera. Brilliant electric lights were placed in tree tops and photographed as the airplane passed over. The camera shutter being open, each light traced a wavy line on the film, and characteristics of which gave an indication of the magnitude and nature of the vibration. Occultation of one of the lights, ten times per second, by a revolving disc introduced the necessary time element.

PROBLEMS

7. The light conditions are such that you wish to give the longest possible exposure while preserving a sharpness corresponding to an image movement of not more than 1/250 inch. When the airplane is flying at an altitude of 10,000 feet, with a ground-speed of 70 miles per hour, what is the longest exposure which can be given if the camera has a 20-inch lens?
8. A series of oblique views are to be taken with a hand camera from a plane flying at an air-speed of 70 miles per hour, at an elevation of 2,000 feet. The camera is held at an angle of 45 degrees to the vertical. It is equipped with a 10-inch lens and carries a 4x5 plate with the 5-inch side horizontal and parallel to the fuselage. What are the longest exposures which can be made without blurring if an image movement of 1/250 inch is permissible? Disregard everything except the ground-speed of the air plane.
9. Under the above circumstances, what depth of territory is covered by the picture from top to bottom?

Figure 3. French Standard 50 cm Camera.

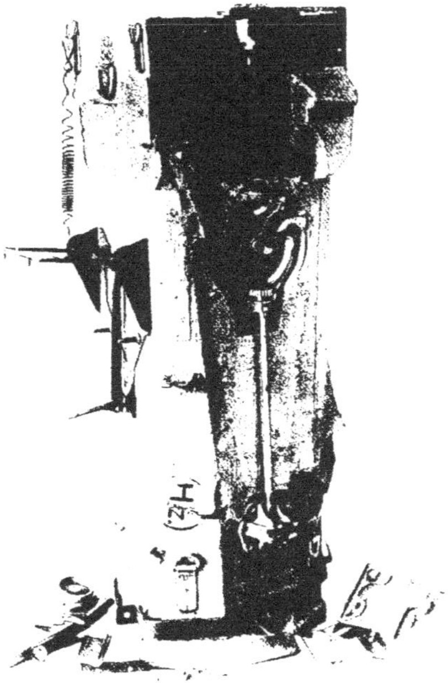

Figure 4. German 50 cm. Airplane Camera. (First enemy camera to be brought down by

10. A camera mounting is being designed for a camera with a 20-inch lens. A sharpness corresponding to an image movement of not more than 1/250 inch is required. What is the maximum permissible angular velocity of the vibration of the camera if it is to be used at a height of 10,000 feet with exposures as long as 1/100 second; (a) When the direction of vibration is at right angles to the motion of the air plane? (b) When parallel? Assume a maximum ground-speed of 90 miles per hour.

11. A camera mounting is being designed for a camera with a 20-inch lens. A sharpness corresponding to an image movement of not more than 1/250 inch is required with exposures of 1/100 second. What is the maximum permissible angular amplitude of vibration if it is known that the frequency of the vibration will be 50 per second; (a) When the direction of vibration is at right angles to the motion of the airplane? (b) When parallel? Assume a maximum ground-speed of 90 miles per hour.

12. The vibration of a camera is often far from simple harmonic. In such cases it is important to know whether it is the amplitude or the angular velocity of the vibration which is the determining factor. Which of the two is the determining factor in the case of a vibration with a frequency of 60 per second, if the time of exposure is 1/50 second? When the frequency is 10 per second and the time of exposure the same?

13. Examination of one of the plates made in studying the effects of camera vibration showed the trace to be an approximate sine wave with a double amplitude of 1/16 inch and a wave length of ¼ inch. There were gaps every ⅛ inch, measured along the line of the average path, corresponding to the occultation of the light every 1/10 second. With this mounting, what is the longest time of exposure which can be given if the image movement must not be more than 1/250 inch?

14. Examination of one of the plates revealed a trace with a maximum angular departure of 45 degrees from the mean direction. The vibration was of such a long period that the maximum angular velocity was the determining factor. The gaps in the trace, corresponding to the occultation of the light, occurred every 0.2 inch long the line of the average path. What is the longest exposure which can be given with this camera if the image motion must not be greater than 1/250 inch?

15. In carrying out the work mentioned above, it was found desirable to do the work in the daytime.

(a) How did the use of red-sensitive plates, a red filter, a small stop, and a brilliant light make possible the carrying out of the experiments by daylight? Discuss the factors separately.

(b) Why were clear glass bulbs employed, rather than red ones?

(c) Show that, other factors being equal, the intensity of the trace obtained in such an experiment varies inversely with the altitude of the camera.

(d) Why was the nature of the background against which the light was placed of importance? What background would you suggest? Give your reason.

16. Refer to the statements made in connection with the last two problems. Assume that the photographic density is proportional to the intensity of the image and the time of exposure and that the extent of woods or dark land is sufficient to form a background for all altitudes and all exposure-times used.

(a) How does the density of the trace vary with the altitude?

(b) Why is the density of the background independent of the altitude?

(c) Show that the density of the trace is independent of its length, but that its contrast with the background is inversely proportional to its length?

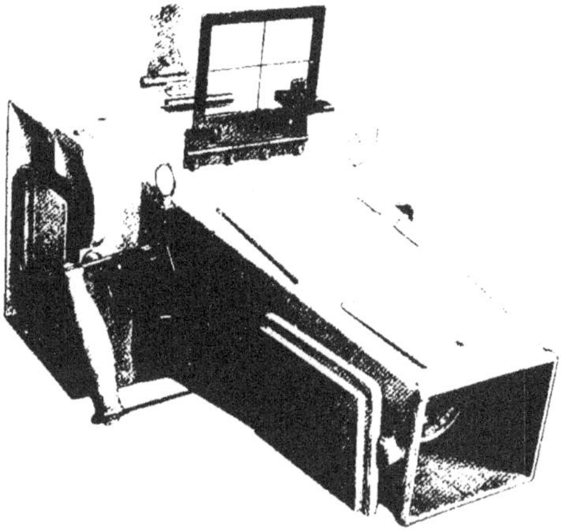

Figure 5. Hand-held Camera Model A-1.
(8 to 10 inch lens. Takes all standard 4x5 Graflex equipment).

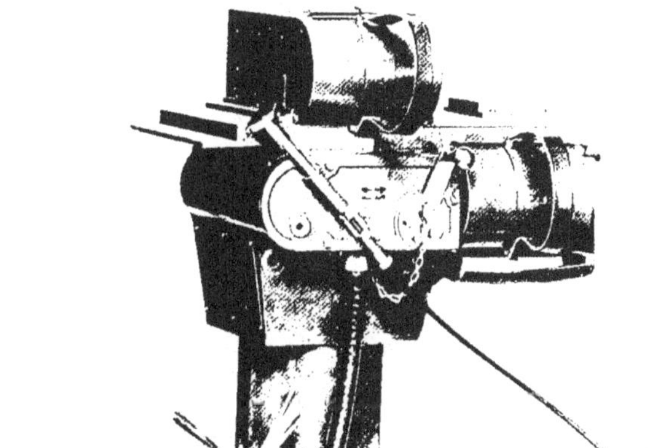

Figure 6. American Type L Camera.
(Copied from Camera most used by British at the front. 8x12 inch lens. 4x5 plates, charges automatically by wind propeller.)

(d) If the length of trace on the plate remains the same, does the contrast with the background vary with the altitude? If so, how?

17. Why do the short wave lengths predominate in the light scattered by "haze"?
18. Why does haze interfere less with the work referred to in Problem 15 than with ordinary aerial photographs taken with regular plates?[1]
19. If haze is to be considered, how will you revise the answers to problems 15 and 16?

---

[1] One of the greatest difficulties encountered in aërial photography is the great amount of haze. The scattered light is predominantly of short wave length so that by the use of yellow, green, orange and even red, screens (made of colored gelatine or glass), combined with plates that are sensitive to the longer wave lengths, it is possible to eliminate to a very large extent the effect of haze. Perhaps the greatest photographic advance produced by the war, was the high commercial development of so-called panchromatic plates, some of which are sensitive even beyond the red end of the visible spectrum. See Part VI, Ray Filters, and Figures 1 and 2, page 130.

PROBLEMS ON CERTAIN RESULTS FROM THE USE OF THE FOCAL PLANE SHUTTER.

20. Answer the following questions in order to ascertain whether or not the distortion resulting from the use of a focal plane shutter is negligible.

(a) A focal plane shutter with a half-inch slit gives an exposure of 1/100 of a second to (each point on) a plate 18 centimeters wide. What is the time of travel of the shutter across the plate?

(b) If the exposure is made from airplane with a ground-speed of 70 miles per hour at a height of 6,000 feet using a 50-centimeter lens, what will be the nature of the distortion; (1) shutter moving in direction of flight? (2) shutter moving in the opposite direction to that of flight?

(c) By measuring on the print, with a scale, the distance between objects whose positions are accurately known, it is found that the scale is accurate when the distances on the print are in a direction perpendicular to the direction of flight. It also appears that the same scale is not accurate for measuring distances, on the same print, in a direction parallel to the direction of flight. What is the greatest possible error, in meters, on the ground? Use data under (a) and (b).

21. If double images are found in successive strips across a plate, taken with a focal plane shutter, and it is expected to eliminate the trouble with future negatives, what is the most probable cause of the difficulty? Why do the double images not occur all over the plate?

22. If a "between-the-lens" shutter were used with the same camera would double images occur? When would they occur? How would they be distributed over the plate?

23. If you are using a focal plane shutter 24 centimeters wide, with a ¾-inch slit, on an 18x24 centimeter plate, for exposures of 1/100 second, in how many places on the plate will there be blurring if the camera is vibrating with a frequency of 50 vibrations per second through an amplitude of 1/10 degree?

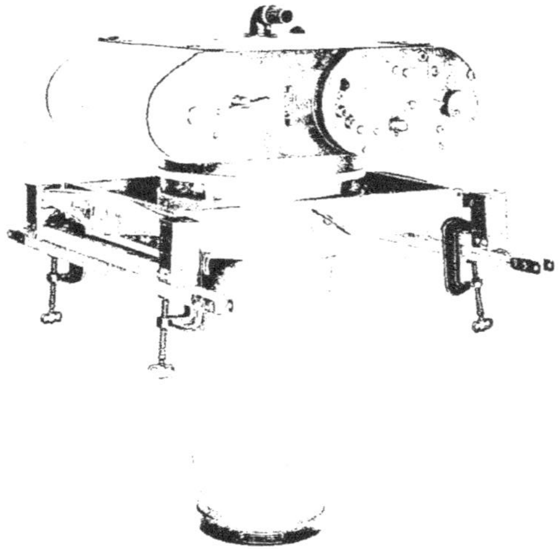

Note the connection for the suction tube and the mounting which permits rotation and tipping in any direction.

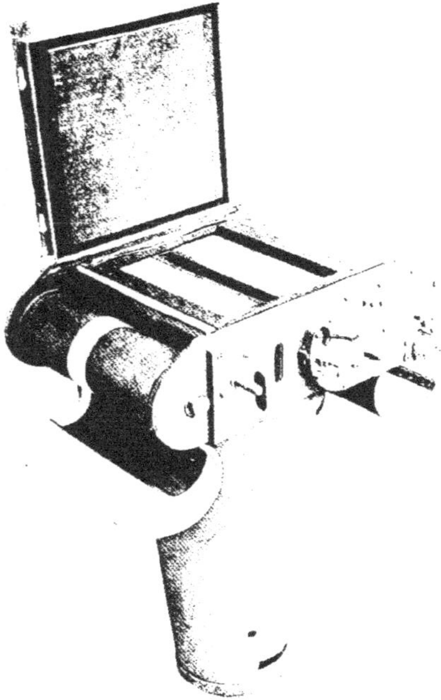

Note focal plane shutter and suction plate, covered with graphited cloth to keep the film flat.

Figure 7. American Type K-1 Fully Automatic Film Camera; (18x24 cm. Film; 100 Exposures; 10 or 20-Inch Lens.)

THE NEGATIVE LENS FINDER

When isolated objectives are to be photographed some sort of sight is necessary. It is especially convenient if the operator can look down through a hole in the bottom of the plane but the construction of the plane seldom permits of a large enough hole. This difficulty was partially overcome by the use of a rectangular negative lens in the floor of the plane.

PROBLEMS

24. A 5x6 rectangular lens with a focal length of minus 11 inches, is to be used as a "finder" for a 18x24 centimeter plate camera with a 20-inch focal length lens. The rectangular lens is to be placed in the floor of the fuselage, 34 inches below the eye of the photographer. What size should a rectangle at the surface of the lens be made if it is to just outline the field of the camera? Note: This lens serves the same purpose as the "direct view finder" of the ordinary hand camera.
25. What is the size of the hole in the floor of the cockpit to give the same angular view as the 5x6 inch lens, when placed with its 5-inch sides parallel to the axes of the fuselage?
26. Where should the lens be placed, flush with the floor or flush with the outside bottom of the fuselage, which may be as much as 5 or 6 inches below the cockpit floor?
27. In bombing it is often very desirable that the bomber should be able to look ahead at a considerable angle through the bottom of the plane. The construction of a plane sometimes makes it impossible for a hole to be cut allowing a forward view beyond 20 degrees. How does the use of a 11-inch negative lens extend the view? By what angle?

PROBLEMS ON THE VENTURI AND SUCTION PLATE OF THE AMERICAN AIRPLANE CAMERA

28. Why is it more important to hold a film flat when using a very rapid lens at full aperture than when using a slow lens? Use a diagram to show this.
29. What is the maximum total pressure on the face of an 18x24 centimeter film if the suction is equivalent to 4 inches of water and there are 289 holes, each 1/64 inch in diameter, in the plate with which the film is backed up?
30. How does a single Venturi tube, like the smaller one in Figure 8, produce suction?
31. How does the outer tube of the double Venturi aid in producing a greater suction?
32. Make a sketch showing how a Venturi could be connected to an aneroid barometer in such a way as to make an air-speed indicator.
33. Make a sketch showing how a Venturi and a Pitot tube could be combined with an aneroid barometer to make an air-speed indicator.
34. Which is better as a surface for the perforated suction plate and as a "wiper" to remove the electric charge, as the film unrolls, a graphited cloth or a sheet of smooth metal? Why? What is the purpose of the graphite?
35. Why are "static marks" more troublesome in winter than in summer?
36. Why are static difficulties experienced with films used in an airplane when no such difficulty is experienced on the same day in using a film camera on the ground?

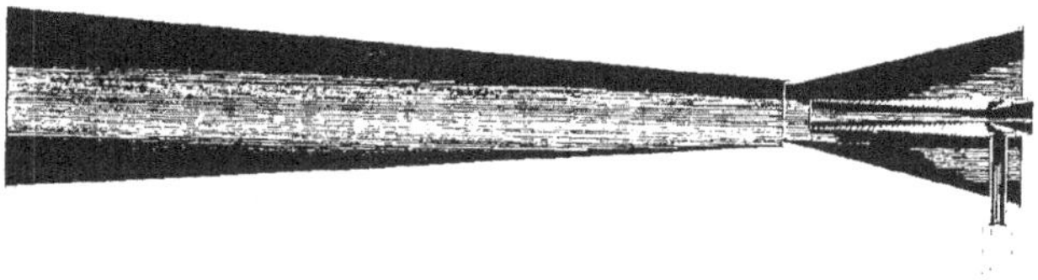

Figure 8. Venturi Tube.
(Produces the suction to hold the film flat in the American type K-1 Camera.)

THE PROBLEM OF THE VERTICAL

In attempting to make maps by aerial photography, it is necessary, if any great accuracy is desired, to take the photographs with the focal plane of the camera horizontal. Therefore, some means of determining the true horizontal or true vertical is of the greatest importance. Such means are much needed for other purposes as well. For example, an artificial horizon will prove of the greatest assistance in aerial navigation and an accurate vertical or horizontal will make possible the accurate dropping of bombs from airplanes. The present methods of sighting are such that no great accuracy can be secured. The natural horizon would, of course, be the simplest way of locating the direction were it not that it is frequently obscured by smoke or haze. A pendulum is of no avail for it will be influenced by every change of velocity.

PROBLEMS

37. An airplane flying at an air-speed of 70 miles per hour in still air makes a perfect bank in making a turn with a radius of curvature of one mile.

(a) What is the ratio of the centrifugal force to that of gravity?

(b) What is the angle between the resultant of these two forces and the perpendicular?

(c) What angle must the wings make with the resultant force if there is to be no side slipping?

(d) At what angle must the plane bank in making the turn?

38. An airplane flying East at an air-speed of 60 miles per hour, in a wind of 60 miles per hour from the North, makes a complete turn to the left.

(a) If it banks at an angle of 10 degrees what is the radius of curvature of the path relative to the air?

(b) Sketch the path relative to the air.

(c) Sketch the path relative to the ground.

39. It is necessary that certain photographs be taken when the camera is pointing not more than two degrees from the vertical.

(a) If the camera is mounted rigidly on the fuselage what is the minimum radius of curvature which the path of the plane can have, in still air, at the instant a photograph is taken?

(b) If the plane is flying in a 60-mile wind is the determining factor the radius of curvature of the path relative to the air or the radius of curvature of the path relative to the ground?

40. An airplane, equipped with a pendulously-mounted camera, is flying at an air-speed of 70 miles per hour in still air. It makes a perfect bank in making a turn with a radius of curvature of one mile.

(a) What is the ratio of the horizontal acceleration to that of gravity?

(b) What is the direction of the resultant force acting on the camera?

(c) What is the position of the camera relative to the plane?

(d) What is the deviation of the camera from the vertical?

(e) Compare this angle with the angle through which the plane is tipped in making the bank.

41. An airplane, equipped with a pendulously-mounted camera, flies East at an air-speed of 60 miles per hour in a wind of 60 miles per hour from the North. It then makes a complete turn to the left.

(a) Sketch the path of the plane relative to the air.

(b) Sketch the path of the plane relative to the ground.

(c) Which path is to be considered in determining the deviation of the camera from the vertical?

(d) At what point in the path does the plane suffer the greatest horizontal acceleration? Give reasons for your answer.

(e) At what point in the path does the camera swing farthest from the perpendicular? Give reasons for your answer.

(f) Why is there a point at which the plane is stationary with reference to the ground?

(g) How does the camera point when the plane is in the position mentioned in (f)?

(h) In what direction, relative to the vertical and the fuselage, does the camera deviate from the vertical just before this point is reached? Just after leaving this point?

42. An airplane is flying 70 miles per hour in a direct cross wind of 15 miles per hour and the pilot attempts to fly a straight course by keeping the nose of the plane headed for an object 15 miles away. How far does a pendulously-mounted camera depart from the true vertical?

43. A camera weighing 90 pounds has a radius of gyration of 12 inches. It is to be hung pendulously in gimbals. How far above the center of gravity must the gimbal bearings be placed in order to give it a double swing period of 10 seconds?

PART V.
BALLOONS

THE U. S. ARMY SPHERICAL BALLOON

The U. S. Army Spherical Balloon consists of a spherical envelope of rubberized cotton or silk containing the gas. The envelope is covered by a net which acts as a suspension for the concentrating ring, to which is attached the basket car to carry pilot and passengers. The valve is at the top of the envelope and is operated from the basket by a cord. The envelope ends in an appendix through which gas escapes as the balloon ascends and air enters as it descends. There are three standard sizes of 9,000, 19,000 and 35,000 cubic feet capacity respectively.

The length of time a spherical or free balloon can remain in the air and the distance it can cover depend largely upon the gas capacity of the envelope and the amount of loose weight or ballast which can be carried.

In order that a spherical or free balloon shall leave the ground with sufficient speed for safety there must be a resultant upward force of from 20 to 100 pounds according to the wind conditions. This is called the ascensional force and is equal to the difference between the "lift" and the total weight of envelope, rigging, car, passengers, etc.

PROBLEMS

1. A spherical balloon is 50 feet in diameter. Assume ground conditions as in the table, page 125.

   (a) What is the weight of the air displaced?

   (b) What is the weight of the gas in the balloon if it is filled with pure hydrogen? If it is filled with illuminating gas?

   (c) What is the total lift in each case?

2. A spherical balloon 50 feet in diameter reaches a height of 10,000 feet. If it is filled with pure hydrogen what is the total lift at this height? What is the lift with illuminating gas? Consult table, page 125.

3. How much gas does a spherical balloon of 35,000 cubic feet capacity lose in rising 20,000 feet? What lift at the ground does this loss represent? Consult table.

4. What is the lift of pure hydrogen in pounds per thousand cubic feet? The lift of illuminating gas? Assume ground conditions as given in the table.

5. What is the lift in pounds per thousand cubic feet of each of these gases at an altitude of 10,000 feet? Consult table.

6. A spherical balloon filled with commercial hydrogen ascends to a point where the aneroid barometer reads 14.5 inches.

(a) At what height is the balloon? Consult table.

(b) What is the approximate temperature? Consult table.

(c) What percentage of the gas has been lost?

(d) If the capacity of the balloon is 50,000 cubic feet what was the total lift at the ground? Assume ground conditions as in table.

(e) What is the lift at the height attained?

(f) What will be the lift when the balloon returns to the ground?

(g) How much gas will have to be valved to just bring the balloon to the ground?

7. The distance which a free balloon can travel depends to a large extent on the ballast which can be carried at the start. Find the relative superiority of commercial hydrogen over illuminating gas for a 35,000 cubic foot spherical balloon in terms of the relative amounts of ballast which can be carried. The balloon and gear weigh 400 pounds, the two passengers weigh 300 pounds. Assume that an ascensional force of 50 pounds is required.

8. A spherical balloon has a capacity of 35,000 cubic feet. The balloon and car weigh 400 pounds, and the two passengers each 150 pounds. How high will the balloon ascend if filled with commercial hydrogen? With illuminating gas? Consult table.

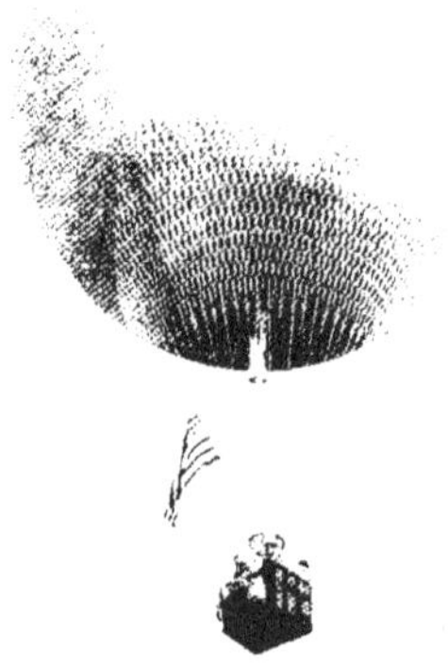

Figure 1. U. S. Army Spherical Balloon.

9. If the balloon and car of a 19,000 cubic foot spherical balloon weigh 300 pounds, and the two passengers each 150 pounds, how high can the balloon ascend if filled with commercial hydrogen? With illuminating gas? Consult table.
10. How high would the above balloon ascend with one passenger?
11. If two free balloons of different volume, both filled with 98 per cent hydrogen, are capable of ascending with exactly the same amount of ballast, which will attain the greater altitude? Why?
12. Why does a spherical balloon "over-shoot" the point of equilibrium, tending to make its flight a series of ascents and descents? This over-shooting means waste of gas or ballast and is prevented as much as possible by a skillful pilot.
13. Why does "over-shooting" result in waste of gas and ballast?
14. What effect have the sun's rays on a balloon? Assume that a free balloon has just enough ascensional force to carry it through a cloud. If the sun heats it 20 degrees Centigrade and it is filled with 35,000 cubic feet of pure hydrogen, how much will the ascensional force be increased? If the top of the cloud is 4,000 feet above the earth how high will the balloon ascend? Disregard the over-shooting.
15. Why is the tension in the fabric in the top of the balloon greater than at the bottom? What is the factor of safety in a 50-foot hydrogen-filled balloon if the fabric tests 38 pounds per running inch, assuming that the fabric does not bulge through the net? Study the photograph, estimate the bulge, and find the approximate factor of safety. What is the tension in the fabric at a point near the appendix?
16. Why is it that a large balloon will lift a considerable weight while a small balloon made from the same fabric will not support even its own weight?
17. Why does the height to which a spherical or free balloon can ascend depend almost entirely upon the size of the balloon?

THE U. S. ARMY OBSERVATION BALLOON

The U. S. Army observation balloon, Type R, is modeled after the French Caquot balloon. It consists of a stream-line envelope with a total length of 90 feet, a maximum diameter of 29 feet, and a gas capacity of 37,000 cubic feet.

In the forward end of the balloon there is a ballonet or air bag much like a huge blister on the inside of the under surface of the envelope. One of the functions of the ballonet is to take care of the variations in the volume of the gas, with change of altitude and temperature. If an observation balloon were open to the air like a spherical balloon, each time it was hauled down a large volume of air would enter the envelope and mingle with the gas. A ballonet makes it possible for air to enter the body of the balloon and not contaminate the gas. When the balloon ascends a second time no gas is lost until all the air is expelled from the ballonet. It is customary to make ballonets of a size to take care of the variations in gas volume encountered under normal conditions and heights of flight.

A second purpose which the ballonet serves is to maintain the pressure in the envelope. The wind often blows against a captive balloon with great force and would cup in the nose were it not that the ballonet automatically maintains the gas pressure sufficiently high to prevent this. The pressure in the ballonet is produced by an air scoop which is placed under the balloon in the zone where the wind pressure on the envelope is the greatest.

The valve is placed in the nose of the balloon and opens automatically when the gas pressure is excessive. It may also be operated by hand.

The type R observation balloon, rigging and basket weigh approximately 1,000 pounds.

PROBLEMS

18. It is common practice to assume the lift of a hydrogen-filled balloon to be 64.4 pounds per thousand cubic feet. Compute the lifting force of hydrogen, at sea level and determine the percentage of purity which is assumed in the case of the above mentioned lift. Consult table, page 125.

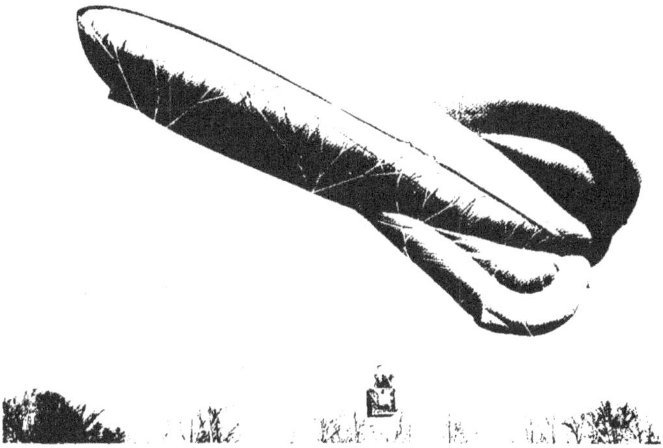

Figure 2. U. S. Army Observation Balloon. Type R.

19. What is the ascensional force in an observation balloon of 37,000 cubic feet capacity? Weight of cordage, car and observer 1,300 pounds, lift of the gas 64.4 pounds per 1,000 cubic feet.

20. What is the ascensional force when the above mentioned balloon has reached 5,000 feet, assuming that the balloon was full when it left the ground? How much gas does a balloon lose in rising 5,000 feet when fully inflated with 37,000 cubic feet of gas?

21. What must be the cubical capacity of a ballonet which will just take care of the variation in gas volume of a hydrogen balloon which is to operate with a maximum elevation of 5,000 feet andwhich has a maximum capacity of 37,000 cubic feet?

22. It is often assumed in rough calculations that a balloon loses 1/8000 of its ascensional force for each meter of altitude. Does this apply to a fully or partially inflated balloon? Consult the table on page 125 and find the maximum error, as a result of this assumption, for altitudes up to 5,000 feet. Up to 10,000 feet.

23. An observation balloon equipped with a ballonet is fully inflated at the ground, the ascensional force is 1,000 kilograms, the cable weighs 1/10 pound per foot. To what altitude will the balloon go?

24. If the same balloon is filled only ¾ full of gas how high will it ascend? A 37,000 cubic foot observation balloon is to ascend to a height of 5,000 feet. With how many cubic feet of 98 per cent hydrogen should it be inflated to secure maximum lift with no loss of gas?

25. The above balloon is to be filled from a number of cylinders containing hydrogen gas at 2,000 pounds pressure at an atmospheric temperature of 80° F. If the actual volume of the tanks is 2 cubic feet how many will be required?

26. A hydrogen-filled observation balloon breaks away at a height of 10,000 meters. The balloon is full and the valve is blocked shut. The pressure at the equator of the balloon is 12 milli-

meters. How high will the balloon ascend before bursting if the greatest diameter is 29 feet and the fabric tests 50 pounds per linear inch?

27. Why is it that an observation balloon, if it breaks away, will turn nose up? Why is this an advantage?

28. A hydrogen-filled observation balloon of 37,000 cubic feet capacity breaks away and rises with a lift force of 600 pounds. If it attains equilibrium at 8,000 feet how much gas has been lost? If more gas is allowed to escape and the balloon comes down, the lower end will become flabby and the balloon will point more and more towards the vertical. Why does this occur?

29. The fabric of an observation balloon with a maximum diameter of 29 feet will stand a tension of 50 pounds per running inch. What is the maximum pressure in inches of water to which the balloon can be safely inflated with hydrogen if a factor of safety of 7½ is to be maintained and the manometer is level with the under surface of the balloon?

30. A winch weighing 5,000 kilograms is placed at right angles to the wind. If the wheels are 2 meters apart and the cable extends at an angle of 45 degrees in a line which intersects the ground at a point half way between the wheels, can the winch be overturned? The cable will stand a tension of 3,200 kilograms.

31. When a winch fails to operate the cable is passed through a snatch block loaded with sand bags. A number of men move along with the block and thus bring the balloon down. If the ascensional force is 600 pounds and the cable is at an angle of 45 degrees, how much must the sand bags weigh to just drag on the ground? How many men must be assigned to the task? (See page 33.)

32. The forces which act on a type R captive balloon on ascension, the air being calm, and the axis of the balloon horizontal, are as follows:

Ascensional force—1,100 kilograms, 11.6 meters from the nose.

Weight of fabric and rigging—400 kilograms, 14 meters from the nose.

Weight of basket and equipment—200 kilograms, 18.6 meters from the nose.

Tension of the cable.

(a) What is the total force acting downward (exclusive of the tension of the cable), the direction of the force, and the distance of its direction line from the nose?

(b) What is the magnitude and direction of the force exerted by the cable and the distance of its direction line from the nose?

33. From what height must a man jump to make the impact equivalent to that of a parachute landing? The rate of descent of parachute is approximately 1,000 feet per minute.

34. The observer lands in a 12 mile wind and the parachute topples over and drags him. With about how much force does the parachute pull him along?

35. A captive balloon is situated 4,000 meters from the enemy lines at the height of 1,500 meters. The wind is blowing directly toward the enemy lines with a velocity of 25 miles per hour. If the observer is forced to jump, where will he land with reference to the enemy lines? A parachute descends at a speed of about 1,000 feet per minute.

36. A captive balloon is situated 6,000 meters behind its own lines at the height of 1,500 meters. If the wind is blowing at an angle of 30 degrees with a line joining the balloon and its own front line trenches, how strong can the wind be and permit the observer to drop in friendly territory? A parachute descends at a speed of about 1,000 feet per minute.

THE FORMATION OF EXPLOSIVE MIXTURES.

Great danger from explosions exists if the hydrogen in a balloon contains more than 18 per cent of air. The following problems serve to call attention to certain causes of explosive mixtures.

PROBLEMS

37. A captive balloon 30 feet high is adjusted so that the pressure recorded by the gauge is 11 millimeters of water. The point of attachment of the gauge and the open end of the gauge are on a level with the middle of the balloon. Calculate the excess pressure in the balloon over atmospheric pressure, (a) at the top, (b) at the bottom of the balloon.

38. Repeat the calculations of Problem 37 for a gauge reading of 3 millimeters of water. What conclusions would you draw in this case as to the effect of making a hole in the lower part of the balloon?

39. Starting with a balloon in the condition represented in Problem 37 trace qualitatively the processes taking place during deflation when, (a) a hole is made in the top of the balloon, (b) a hole is made in the bottom of the balloon, (c) one hole is made at the top and another hole at the bottom of the balloon.

40. A captive balloon 29 feet high is filled with hydrogen until a gauge at the middle reads 11 millimeters of water. The valve is adjusted to open at 15 millimeters pressure. The balloon ascends to an altitude of 3,000 feet and returns to the ground. Calculate the minimum volume which must be provided in the ballonet in order that the pressure at the bottom of the balloon shall not be less than that just outside when the balloon returns to earth.

41. A balloon is filled as in Problem 40. The gauge is then placed in the basket 50 feet below the point of attachment to the balloon. What is the change in reading of the gauge? How would the result be effected if the tube leading from the gauge to the balloon was filled with air?

42. A balloon of 37,000 cubic feet capacity has a hole 1/10 of an inch in diameter in its side where the excess pressure is 11 millimeters. Calculate the fall in pressure per day as a result of this hole.

43. The hole mentioned in Problem 42 might easily happen to open into a comparatively closed pocket like the fins, of say 15 cubic feet capacity. About how long a time would elapse before the mixture in the closed pocket became explosive. A mixture containing more than 18 per cent of hydrogen is explosive.

44. How does pressure difference affect the passage of gas through a hole in the fabric? Do the same considerations apply to pure diffusion? Explain.

45. Assuming that 9 liters of hydrogen pass through one square meter of fabric by diffusion in 24 hours, at what rate would you expect air to enter the balloon owing to diffusion? How long would it take for diffusion to result in an impurity of 18 per cent of air in a spherical balloon of 37,000 cubic feet capacity?

THE U. S. ARMY 84,000 CU. FT. AIRSHIP

The U. S. Army 84,000 cubic foot airship or dirigible is illustrated in Figure 3. This airship is of the non-rigid type, the shape of the envelope being maintained by the gas pressure, which is always greater than that of the atmosphere. When the airship is under way a pressure of at least 18 millimeters is maintained. If the pressure reaches 38 millimeters, approximately, gas is valved automatically.

Two ballonets [1] are used; one in the nose and one near the tail. These are connected with a small blower and also with a "scoop" placed in the slip stream of the propeller. By either of these means pressure can be maintained in the ballonets. This pressure is communicated to the gas and must be sufficient to maintain the shape of the balloon against the wind pressure.

The ballonets are also employed in altitude control. Valves are arranged so that either ballonet can be inflated at will. Transferring 1,000 cubic feet of air from one to the other tilts the balloon 5 degrees. When the balloon is under way this furnishes a very effective means for vertical steering, making the use of the elevators unnecessary and doing away with the resistance attending their use. A skillful pilot will depend almost entirely upon the ballonets.

---

[1] See page 112.

Figure 3. U. S. Army 84,000 Cubic-Foot Airship.

PROBLEMS

46. How much lift at the ground would an 84,000 cubic foot airship have if inflated with pure hydrogen? With pure helium?
47. How much would the lift be reduced for each cubic foot of air in the ballonets, (a) if the airship were inflated with pure hydrogen, (b) with pure helium?
48. What is the lift at the ground of the Army 84,000 cubic foot airship when the gas is commercial hydrogen and the ballonets are full? When commercial helium is used and the ballonets are full?
49. How high can an 84,000 cubic foot airship, filled with commercial hydrogen, ascend without losing gas if it valves at 38 millimeters and the manometer reads 18 millimeters when the airship leaves the ground, with ballonets full?
50. How high can an 84,000 cubic foot airship, filled with commercial helium, ascend without losing gas if it valves at 38 millimeters and the manometer reads 18 millimeters when the airship leaves the ground, with ballonets full?
51. The 84,000 cubic foot airship is to fly at altitudes not to exceed 5,000 feet. Maximum lift is desired with no loss of gas. How much air should be left in the ballonets if the envelope is inflated with commercial hydrogen at a pressure of 18 millimeters and the pressure is not to exceed 30 millimeters at the maximum elevation of 5,000 feet?
52. What will be the answer to problem 51 if helium is substituted for hydrogen?
53. How many cylinders of hydrogen gas will be required to fully inflate the Army airship when the atmospheric pressure is 740 millimeters and the temperature is 10° C.? The cylinders were filled with 6.4 cubic meters of free gas at a pressure of 760 millimeters and temperature of 15° C.
54. Some gas cylinders are filled with gas at a pressure of 150 kilograms per square centimeter at a temperature of 15° C. If at this pressure the factor of safety is 5, is there danger of bursting when the temperature of the cylinder is raised 40 degrees Centigrade by the sun?

55. How much more does it cost to completely inflate the 84,000 cubic foot airship with helium than with hydrogen? Helium costs $100 per 1,000 cubic feet and hydrogen $8.

56. How much less lift has commercial helium than commercial hydrogen? Express in percentage of hydrogen lift.

57. In view of the facts brought out in the answers to the last two problems, why should it be proposed to use helium in all captive balloons and airships?

58. A mixture of helium and hydrogen containing 15 per cent of the latter will not ignite.

(a) What is the relative lifting power of commercial helium, commercial hydrogen, and the mixture? Assume the density of impurities to be that of air.

(b) If the same lift which is secured with 1,000 cubic feet of commercial helium is obtained by the use of the mixture what is the saving in cost?

59. Helium is obtained by liquifying and distilling off all the other gaseous constituents of a natural gas obtained in the Southwest. The helium is obtained in a practically pure state. The other gases are nitrogen, methane, ethane, propane, butane and pentane. What are the physical characteristics of these other gases and of helium which make this process possible?

60. When ballonets are used in a balloon they are usually made of such a capacity as to become deflated at the maximum height at which the balloon is to be used. What percentage of the total balloon capacity must be alloted to the ballonet in the case of an observation balloon to operate at a maximum height of 1,500 meters?

61. The Italian Type M airship is designed for a height of 18,000 feet on account of the high mountains to the North. Why are very large ballonets employed? About what percentage of the total balloon capacity must be alloted to the ballonet? Assume a maximum pressure of 38 millimeters and that the pressure may reduce to zero when the ship returns to the ground. This latter condition is made possible by the fact that the Italian ships are semi-rigid.

62. Why will a submarine, not under way, either float to the surface or sink to the bottom while a balloon will, if it ascends at all, tend to reach a certain level at which it will float steadily?
63. Why can the observer in a free balloon locate the point on the ground immediately below him by sighting with the trail rope, while a trail rope or its equivalent will be of no use in a captive balloon, an airplane, or a dirigible?
64. Give reasons why the cars of a dirigible should be placed some little distance in front of the center of lift.
65. Approximately how many pounds are added to the gross lift of the 84,000 cubic foot airship for each drop of temperature of 10 degrees Fahrenheit?
66. The instructions regarding the operation of the dirigible are not to start after sunrise with less than 100 pounds of quickly available ballast and that double the amount will usually be necessary in the late afternoon. Why is more ballast needed in the afternoon than in the morning?
67. Why does every 1,000 feet gain in altitude with the bag full of gas mean a loss of lift of about 160 pounds? Why must this equivalent of load be discharged before landing?
68. Why are instructions given to never force a dirigible up higher by power than it would go by use of ballast?
69. How near does the consumption of fuel, etc., keep place with the loss of lift if the leakage amounts to 3,000 cubic feet per day?

NAVIGATION CHART

A navigation chart showing the net speed that can be made in any direction under different wind conditions is of great assistance in the proper navigation of airships and airplanes. The normal speed of the machine is taken as 1. The angles between the course of the wind and the direction of the destination are plotted horizontally, and the resulting net speed of the airship is plotted vertically. The wind velocity is expressed in terms of the air-speed of the machine. Wind velocities equal to 2, 1½, 1, ½, 1/3, and 0, times the air-speed are convenient values. Some students may find it interesting to prepare a complete chart. Certain of the following problems can be answered without the use of the chart.

70. Prepare a navigation chart as described above.
71. Can a line be drawn on the chart which represents the maximum speed which can be attained with any wind, no matter how high? If so, draw the line on the chart.
72. Why does the ordinate of this curve for any angle give the wind which will give maximum speed for that particular angle?
73. Under what conditions is a wind blowing at more than 90 degrees to the course unfavorable? Is a wind at less than 90 degrees always favorable? Explain.
74. Why is it that a wind of very high velocity, compared to that of the ship, is always unfavorable except when the angle is zero?
75. Why do curves for 1½ and 2 times the air speed of the ship suddenly end?
76. Within how many degrees of the course must the wind be blowing to be favorable if the wind velocity is equal to the air speed of the ship?
77. If the wind velocity equals the air speed of the ship can the ship be held to its course for wind directions from 90 to 180 degrees from the course? Will there be any progress?

DATA ON THE 84,000 CU. FT. AIRSHIP

1. Linear—

Length	163 ft.
Mean diameter of maximum section	31½ ft.
Maximum width (across fins)	41 ft.
Maximum height	46 ft.
Length of car	28 ft.

2. Volumes—

Envelope (total)	84,000 cu. ft.
Forward ballonet	8,500 cu. ft.
Rear ballonet	12,500 cu. ft.

3. Total Fin Surfaces—

Fixed horizontal	304 sq. ft.
Elevators	70 sq. ft.
Fixed vertical	152 sq. ft.
Rudder	56 sq. ft.

4. Power Plant—

Curtiss OXX-3 engine, N-9 pusher, pro-propeller, 8′ x 4′-7″, direct connected. 1,415 r. p. m. at full speed. Maximum brake horse-power, 105.

5. Weights—

Gas bag proper	1,550 lbs.
Gas bag and associated parts	1,760 lbs.
Fin surfaces	310 lbs.
Car, engine, tanks, etc	1,280 lbs.
Instruments, parachutes, tools, etc	350 lbs.
Two men	320 lbs.
Necessary ballast	100 lbs.
Fuel and oil (full)	650 lbs.
Gas, 84,000 cu. ft. of hydrogen (97 per cent quality)	630 lbs.

6. Gross Displacement at 70° F. and 30″ Barometer .... 6,300 lbs.

7. Speed and Resistance—

Maximum air speed....................	48 m. p. h.
Minimum resistance at 48 m. p. h........	540 lbs.
Resistance at other speeds..............	R=.33 $^{1.9}$
Ordinary cruising speed................	39 m. p. h.

Vertical resistance of balloon without power, 3.5 S^2, in which S is the vertical speed in feet per second.

8. Variations in Lift and Load—

Leakage of gas............	2,000 to 5,000 cu. ft per day
Fuel consumption at full power......	65 lbs. per hour
Fuel consumption at cruising speed..	46 lbs. per hour

Evening contraction may produce a surplus load of as much as 400 lbs.

Rain may add a load of 400 lbs.

Every 1,000 feet altitude with the bag full of gas means a loss of lift of about 160 lbs.

AVERAGE AIR TEMPERATURES AND DENSITIES AT DIFFERENT ALTITUDES.[1]

(Standard density=1.293 Kg. per cu. m., 0°C., 76 cm. pressure.)

Altitude	Pressure	Temperature	Vapor pressure	Atmospheric density	
				Per cent standard	Per cent surface
Feet	*Inches*	°*F*	*Inches*		
0	29.92	59.0	0.36	94.4	100.0
1,000	28.86	56.0	.32	91.6	97.1
2,000	27.83	52.9	.28	88.9	94.2
3,000	26.83	49.8	.25	86.2	91.4
4,000	25.86	47.3	.22	83.6	88.6
5,000	24.92	44.9	.19	80.9	85.8
6,000	24.02	42.5	.16	78.4	83.1
7,000	23.12	40.0	.14	75.9	80.4
8,000	22.28	37.3	.12	73.5	77.9
9,000	21.46	34.3	.105	71.3	75.5
10,000	20.66	31.2	.086	69.1	73.2
11,000	19.88	27.9	.072	66.9	70.9
12,000	19.13	24.6	.060	64.8	68.7
13,000	18.40	21.3	.050	62.8	66.6
14,000	17.69	17.8	.045	60.8	64.5
15,000	17.00	14.2	.040	58.9	62.4
16,000	16.35	10.8	.035	57.1	60.5
17,000	15.72	7.4	.030	55.3	58.6
18,000	15.10	4.0	.025	53.5	56.7
19,000	14.50	0.4	.020	51.8	54.9
20,000	13.92	— 3.4	.017	50.1	53.1
21,000	13.36	— 7.2	.015	48.5	51.4
22,000	12.82	—11.0	.013	47.0	49.8
23,000	12.29	—14.8	.010	45.4	48.1
24,000	11.79	—18.4	.008	43.9	46.5
25,000	11.30	—22.0	.007	42.4	45.0
26,000	10.83	—25.6	.006	41.0	43.5
27,000	10.37	—29.2	.005	39.6	42.0
28,000	9.93	—32.6	.004	38.2	40.5
29,000	9.51	—36.0	.003	36.9	39.1
30,000	9.10	—39.2	.002	35.4	37.7

DENSITY OF GASES, 0°C., 76 CM. PRESSURE.

Hydrogen .... .0056 pounds per cu. ft.
Helium .... .0112 pounds per cu. ft.
Illuminating gas (approximately) .... .0500 pounds per cu. ft.
Air .... .0807 pounds per cu. ft.

(Commercial hydrogen is 98 per cent pure, helium 92 per cent. In all problems assume impurities to have the density of air.)

[1] From the Monthly Weather Review, March, 1919, 47 : 157.

PART VI.

RAY FILTERS

RAY FILTERS

The information necessary for the answering of the following questions can be found in any good optical text. Preston's "Theory of Light" or Wood's "Physical Optics" are suggested as suitable books. It is believed that an instructor will have no particular difficulty in using the following questions in an elementary course if a little time is devoted to furnishing the necessary background. The use of ray filters in photography is also treated under the subject of Aerial Photography. Figures 1 and 2, show how successfully the effects of "haze" have been eliminated.

In order to illustrate the subject matter of the last four questions, the instructor should have at hand sample filters. Small discs of cobalt blue glass and of the various noviol glasses manufactured by the Corning Glass Works, Corning, Pa., are recommended. By combining the blue glass with the different noviol glasses, very interesting filters are produced. The most striking results are obtained when different green pigments, apparently the same to the eye, are examined through these filters.

PROBLEMS

1. Why does the light scattered by fine particles in the air interfere with the distinctness with which distant objects can be seen?
2. What particular color or portion of spectrum would you expect to predominate in the scattered light?
3. Can this scattered light be eliminated by means of a suitable ray filter?
4. Characterize roughly the spectral transmission which would seem desirable in a ray filter for observing distant objects.
5. Why will distant objects appear more distinct when viewed through such a ray filter?
6. What will be the color of such a ray filter when viewed by transmitted light?

7. State in general terms how the apparent color of red objects will be affected. Of blue objects.
8. In photographing a distant view why is a ray filter desirable?
9. Why is it that the common yellow ray filter used on a camera to secure "cloud effects" by reducing the intensity of the blue light of the sky would be useful in securing better definition of distant objects?
10. What are the optical characteristics of such a ray filter?
11. Why should the relative sensitivity of the plate for different portions of the spectrum be taken into account in order to determine whether the spectral transmission of the ray filter is satisfactory?
12. State briefly the optical characteristics of a ray filter to be used with orthochromatic plates, which are sensitive well down into the red end of the spectrum.
13. What different characteristics would you expect in a filter to be used with ordinary plates, the maximum sensitivity of which lies in the blue or even in the ultra-violet portion of the spectrum?
14. Suppose the filter designed for ordinary plates is used for both ordinary and orthochromatic plates, which would require the greater increase in exposure?
15. Why should one take into account the visibility curve of the eye in selecting a filter to be used for visual observation?
16. State briefly what modifications might be made in the requirements to be satisfied by a filter when the visibility curve is taken into consideration, as compared with those based upon the assumption of uniform sensitivity throughout the visible spectrum.
17. In the above questions it has been tacitly assumed that the sun is the source of illumination. Suppose a searchlight is to be used at night. In what part of the spectrum will the scattered light be most abundant?
18. What can you say of the relative advantages of viewing the distant object from a point near the searchlight and from a point some distance to one side of the searchlight?

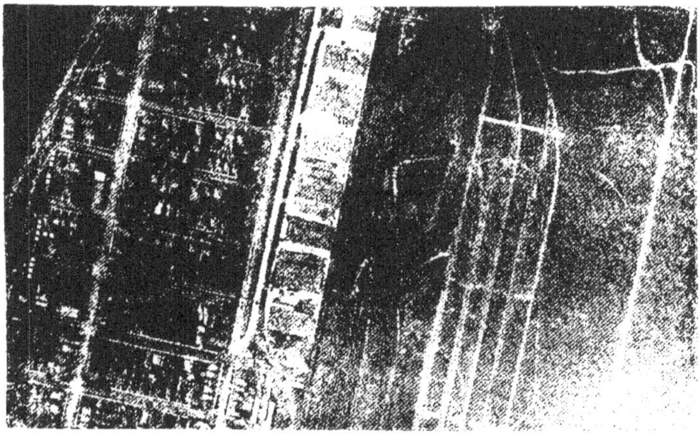

Figure 1. Area near Langley Field. Va. Photographed with an ordinary plate, with no filter.

(Exposure 1/340 second; altitude 5,500 feet. Low altitude gives an advantage over plate used for photograph below.)

Figure 2. Same area photographed with a special American Panchromatic plate, using a red filter.

(Exposure 1/30 second; altitude 10,000 feet.)

19. What are the optical characteristics of a ray filter to be used when viewing a distant object illuminated by a searchlight?
20. How could one modify the glass in front of a searchlight so as to secure the same result as is obtained with a ray filter in front of the eye?
21. What characteristics should be given to the arc in order to secure this same result?
22. What characteristics could be given to the reflector in order to secure this same result?
23. The French did, in fact, substitute gold for silver on their searchlight mirrors. How does the effect produced by the gold harmonize with the characteristics which would be desirable in a reflector?
24. This gold reflecting surface worked well when the searchlights were used for the observation of terrestrial objects, but when used for anti-aircraft work, the enemy quickly appreciated the lack of a portion of the spectrum and painted their airplanes so that they were invisible when illuminated with these searchlights. What color did they paint the airplanes?
25. Buildings surrounded by green foliage are frequently painted green in order to lessen the contract between them and the trees. When viewed through an appropriate filter, the foliage may appear black while the painted surfaces appear red. What does this seem to show regarding the relative amounts of red and green light reflected by green foliage and the pigment of the paint?
26. With other painted surfaces and different filters the foliage may appear red and the paint black. What does this show?
27. What can one say of the transmission of the filter in each case?
28. In general what should be the characteristics of a filter designed to increase contrast between the apparently similar colors?
29. Describe in general terms how signalling can be accomplished by the use of a light which appears practically uniform in intensity to the enemy, but is seen by the friendly observer as a series of flashes.

PART VII.

OPTICAL SYSTEMS OF INSTRUMENTS

OPTICAL SYSTEMS OF INSTRUMENTS

An attempt has been made in the following pages to furnish material which will be of assistance to the instructor who wishes to introduce into a course in Light, subject-matter dealing with the simpler optical systems used in ordinary optical instruments and in military instruments. Much of the material is suited to an elementary course. The following books are mentioned as references:

Bureau of Standards, "The Properties and Testing of Optical Instruments, Circular No. 27."

Drude, P. K. L., "The Theory of Optics," Translated by C. R. Mann and R. A. Millikan, Longmans Green & Co.

Gleichen, Alexander, "The Theory of Modern Optical Instruments," Translated by H. H. Emsley and W. Swaine. His Majesty's Stationery Office, London.

Nutting, P. G., "Outlines of Applied Optics," P. Blakiston Son & Co., Philadelphia, Pa.

Southall, T. P. C., "Mirrors, Prisms and Lenses," The Macmillan Company, New York, N. Y.

The instructor not already familiar with military instruments will find the book by Gleichen indispensable. It was translated by the British Government in order to supply information to Army officers and manufacturers of Army instruments and contains two chapters on range finders which constitute one of the best sources in English for information in regard to this instrument. It also contains a chapter on other military instruments as, for example, the panoramic sight and the battery commander's telescope.

No attempt has been made to give an extended description of the different instruments or of their use. However, there are furnished diagrams of the optical systems of the instruments and of the different types of prisms. These will be found useful in connection with the problems. The student will have to depend upon reference books and the instructor for the necessary background.

It is felt that wherever possible the actual instruments and prisms should be placed in the hands of the students. It will be difficult for an instructor to adequately present to a class the material dealt with in the questions unless he has at hand the actual prisms and instruments as outlined above. The list is as follows[1]:

Opera glass to illustrate the Galilean telescope.
An inverting telescope.
A terrestrial telescope with lens erecting system.
Right angle telescope.[2]
A prism binocular to illustrate the Porro erecting system.
Right angle prism.
Roof prism.
Penta prism.
Porro prism system.
Battery commander's telescope.
Panoramic sight.
An 80-centimeter or 100-centimeter horizontal base range finder.

Set of panoramic sight optics mounted in skeleton form so that the system forms a practicable telescope. The objective prism and rotating prism should be mounted in such a way that they can be rotated, but it is not necessary to have them connected by gears as in the completed instrument.

[1] An institution having a Reserve Officers' Training Corps Unit should enlist the assistance of the Commandant in procuring as much of this equipment as possible.

[2] If desired this may be omitted as the lower half of the panoramic sight is a complete right angled telescope.

Figure 1. Panoramic Sight.

Figure 2

THE GALILEAN TELESCOPE

1. Locate the image formed by the objective of the telescope.
2. Trace the course of the rays through the instrument showing that the final image is erect. Is this final image real or virtual?
3. When the adjustment is "telescopic," i. e., when the focal planes of eyepiece and objective coincide, and the entering pencil is parallel, is the emergent pencil parallel or convergent?
4. Locate the image formed by the telescope (not the objective) when adjustment is "telescopic."
5. Is the image formed by the telescope any nearer the eye than the object viewed?
6. How then does the telescope magnify?
7. Give the formula for the magnification.
8. Is there any difference in the "accommodation" of the eye when viewing a distant object and viewing the image through the telescope in "telescopic" adjustment?
9. If one is near sighted, in what direction must the ocular, in telescopic adjustment, be moved for comfortable vision?
10. Locate the exit pupil of the Galilean telescope.
11. Why cannot the pupil of the eye be brough to coincide with the exit pupil in this type of telescope?
12. What effect does this fact have upon the field of view?
13. What effect upon the illumination of the field?

14. Is there any plane within the instrument in which an object could be placed and appear sharp to the observer's eye when the eye is accommodated for the distant object under observation?
15. What bearing does this have upon the place at which a reticule, having cross-lines, should be put?
16. What bearing does this have upon the place at which a diaphragm should be placed in order to provide a sharp well-defined boundary for the field of view?
17. Assuming that the maximum field of view for the eye is 45 degrees, what is the maximum actual field for a six-power instrument?
18. What is the relation existing between the exit pupil, entrance pupil, and magnification?
19. Assuming that the exit pupil is to be 6 millimeters in diameter, what must be the diameter of the objective of a six-power instrument?
20. What will be the maximum resolving power of the objective?
21. What will be apparent value of the angle subtended by two points just resolved by the objective when viewed through the six-power binocular?

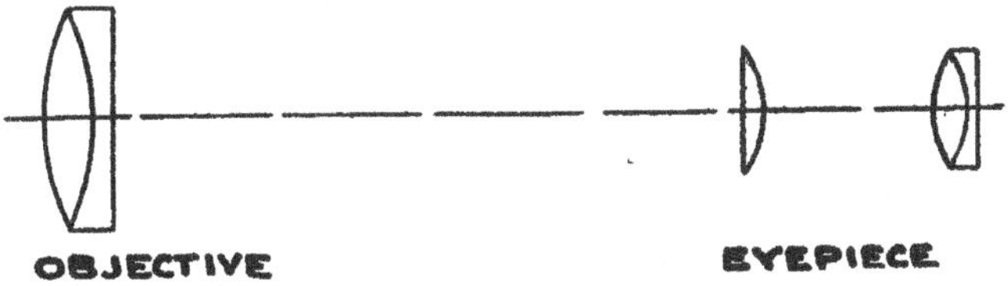

Figure 3

INVERTING OR ASTRONOMICAL TELESCOPE

22. Locate the image formed by the objective.
23. Where is the plane within the instrument which is in focus for the observer's eye when the eye is accommodated for the distant object?
24. Where should a recticule having cross lines or scale be placed?
25. Why do the cross lines appear stationary with respect to the object when the eye is moved?
26. State clearly what is meant by parallax.
27. Why is there parallax when the reticule is displaced from its normal position?
28. If, as the eye is moved from side to side in front of the eyepiece, the cross lines appear to move, relative to the distant object, in the same direction with the eye, is the reticule nearer or farther from the objective than the focal plane?
29. Where in the instrument should a diaphragm be located in order to provide a sharp boundary for the field?
30. What must be the diameter of this diaphragm if the focal length of the objective is ten inches and the field is four degrees?
31. Locate the exit pupil of this telescope.
32. Why can the pupil of the eye be brought into coincidence with the exit pupil in this type of telescope?
33. What effect does this fact have upon the field of view?
34. What effect upon the illumination of the field?
35. What is the formula for the magnification of this instrument?

36. Assuming that the exit pupil is as large as the pupil of the eye, is the brightness of the image increased by enlarging the objective?
37. If you believe that the brightness is not increased, draw a diagram showing clearly why the extra area of the objective does not contribute to the illumination of the image.
38. If two telescopes have exit pupils of the same size, upon what may a difference of illumination obtained with the two instruments depend?
39. Why does an increase in the size of a photographic aperture always increase the brightness of the image?
40. Wherein is the difference between this case and that of enlarging the objective of the telescope?

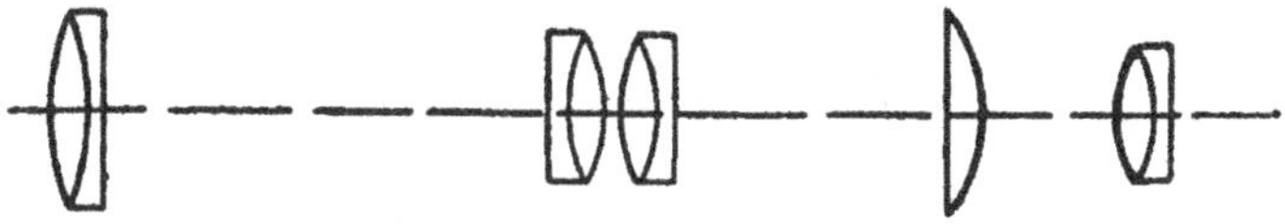

Figure 4

TELESCOPE WITH LENS ERECTING SYSTEM

41. Trace rays through this system and show that the final image is erect.
42. Locate two planes, each of which is in focus when the observer's eye is accommodated for the distant object?
43. Why can a reticule be placed in either of these planes? Should a reticule with a marked scale be placed in the instrument as it is to appear or should it be reversed or turned upside down or both?
44. In what place could a diaphragm be placed to sharply limit the field of view?
45. How does the position of the erecting system influence the magnification produced by the instrument?
46. Develop a formula for the magnification of a telescope with a lens erecting system.
47. Locate the exit pupil of this telescope.

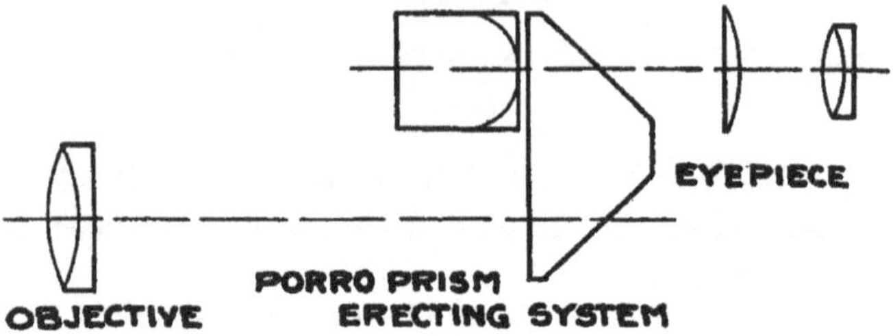

Figure 5

TELESCOPE WITH PORRO PRISM ERECTING SYSTEM

48. Trace the rays through a pair of porro prisms. Is the image erected?

49. How many reflections do the rays undergo?

Figure 6. One Meter Horizontal Base Range Finder.

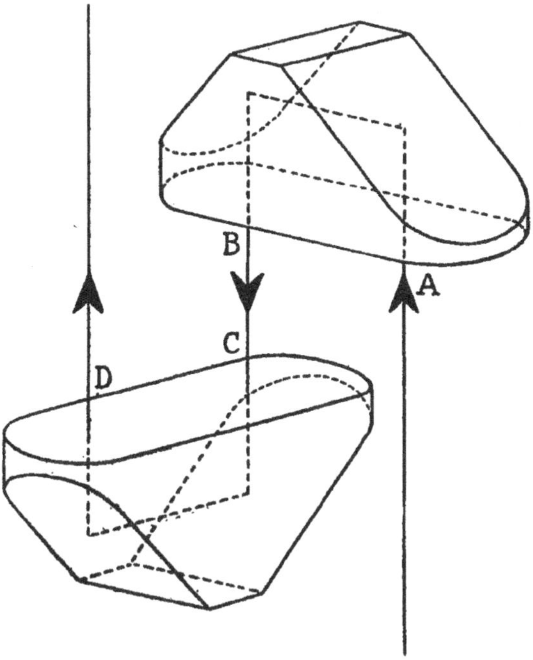

Figure 7. Porro Prism

PRISMS

50. If the reflection of a printed page is viewed in a mirror, the letters are observed to be reversed. This is commonly spoken of as perversion of the image. Is the image perverted after a single reflection in a prism?
51. Is it perverted after two reflection?
52. Is the image perverted after an odd number of reflections?
53. Is the image perverted after an even number?
54. Is the real image formed by a lens perverted?
55. Is a virtual image formed by a lens perverted?
56. Does any combination of lenses ever pervert an image?

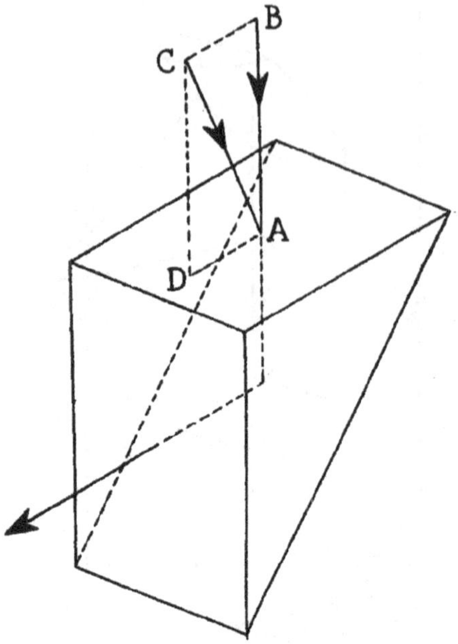

Figure 8. Right Angle Prism

57. Is the image formed by a telescope with a lens erecting system perverted?
58. The erection of the image is to be accomplished by means of prisms. Should an even or an odd number of reflections be employed, if the image is to be unperverted?
59. If a telescope gives an erect and unperverted image, in what direction does the field move when the telescope is turned to the right?
60. Is the relationship between movement of field and movement of telescope a natural one?
61. If the telescope gives an image erect, but perverted, in what direction does the field move when the telescope is moved to the right?

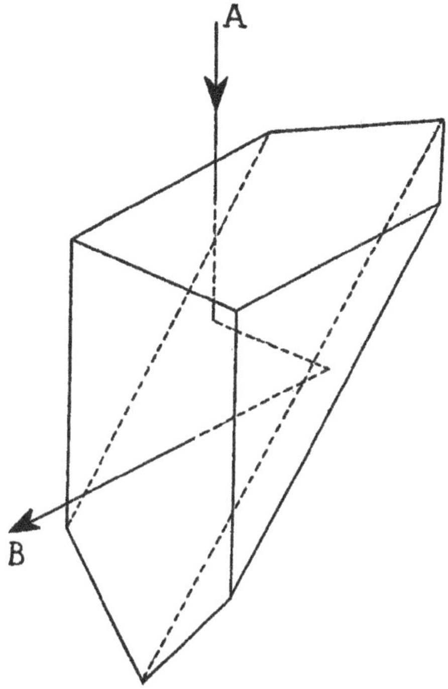

Figure 9. Roof Prism

62. Is the relationship between movement of field and movement of telescope a natural one?
63. Which type of telescope would be more advantageous for laying a gun, that of Problem 59 or 61?
64. In general, what statement may we make regarding the number of reflections which light must undergo in a military telescope?
65. In the drawing of a right angle prism, BA is a ray incident normally upon the face of the prism and emerging after a reflection on the hypothenus face. Will there be total reflection if the index of the glass is 1.52?

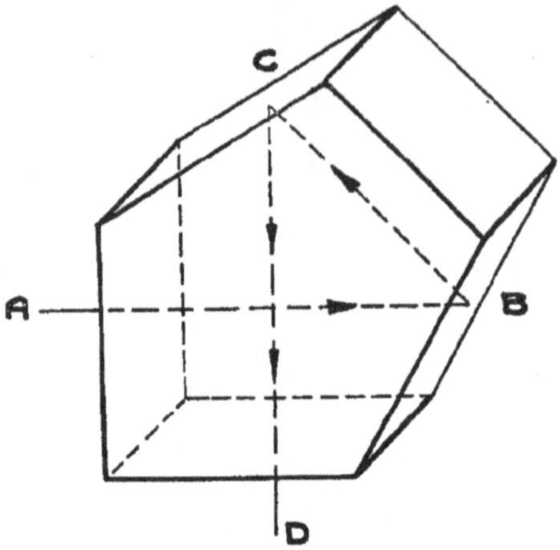

Figure 10. Penta Prism

66. CA is a second incident ray (not traced through the prism) lying in the plane ABCD and making an angle CAB with the ray BA. What is the maximum value of angle CAB, if total reflection is to be secured at the hypothenus face?
67. Suppose we have any other ray, incident at A, making an angle with BA less than or equal to this maximum value, but not lying in the plane ABCD, will it be totally reflected?
68. What limit does this set upon the maximum angular value of the field which may be viewed, fully illuminated, by means of an unsilvered right angle prism?
69. What is the limit for a prism made of glass with an index of reflection of 1.52?
70. Will this angular extent of field be increased if flint glass, with an index of refraction of 1.62, is used?
71. What will the angular extent of field be for 1.62 glass?
72. Can the field of a right angle prism be increased by silvering the hypothenuse face?

73. The roof prism bends a ray, incident normally upon one face, through 90 degrees, as does the right angle prism. However, in the roof prism there are two reflections, whereas in the right angle prism there is only one. What can be said of the difference between the image formed by the one prism and that formed by the other?

74. In the penta prism the reflection takes place at two faces inclined to each other at 45 degrees. What is the total deviation of the ray?

75. Is the image shifted when the penta prism is rotated slightly about an axis parallel to the intersection of the two reflecting faces?

76. Is the deviation of the rays altered when the prism is rotated?

77. Is the image viewed in the penta prism perverted?

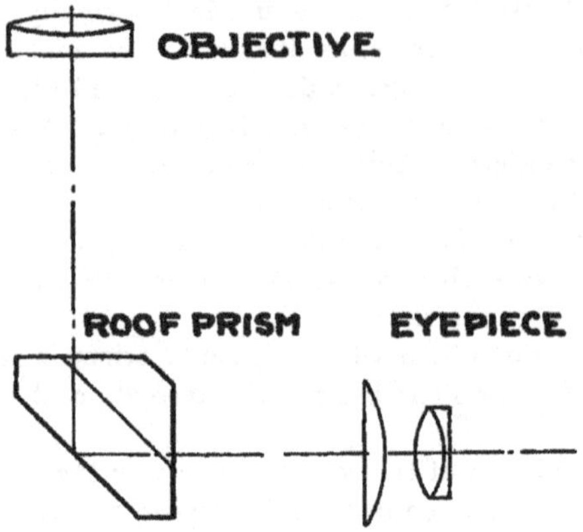

Figure 11

RIGHT ANGLE TELESCOPE

78. In the right angle telescope shown, the prism used is a roof prism. Why is the roof prism required instead of an ordinary right angle prism?

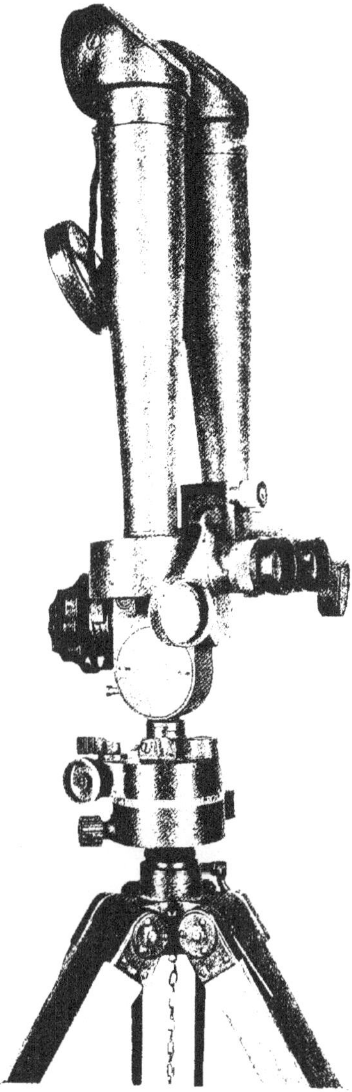

igure 12. Battery Commander's Telescope.

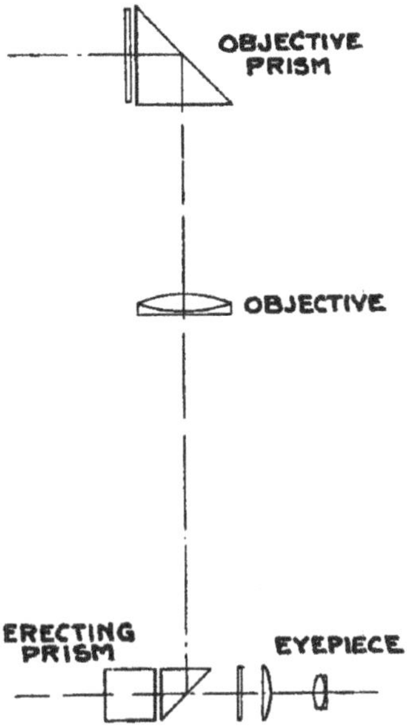

Figure 13

BATTERY COMMANDER'S TELESCOPE

79. Trace a pair of rays through the Battery Commander's telescope and discover whether the image is erect or not.
80. Count the number of reflections and determine whether the image is perverted.

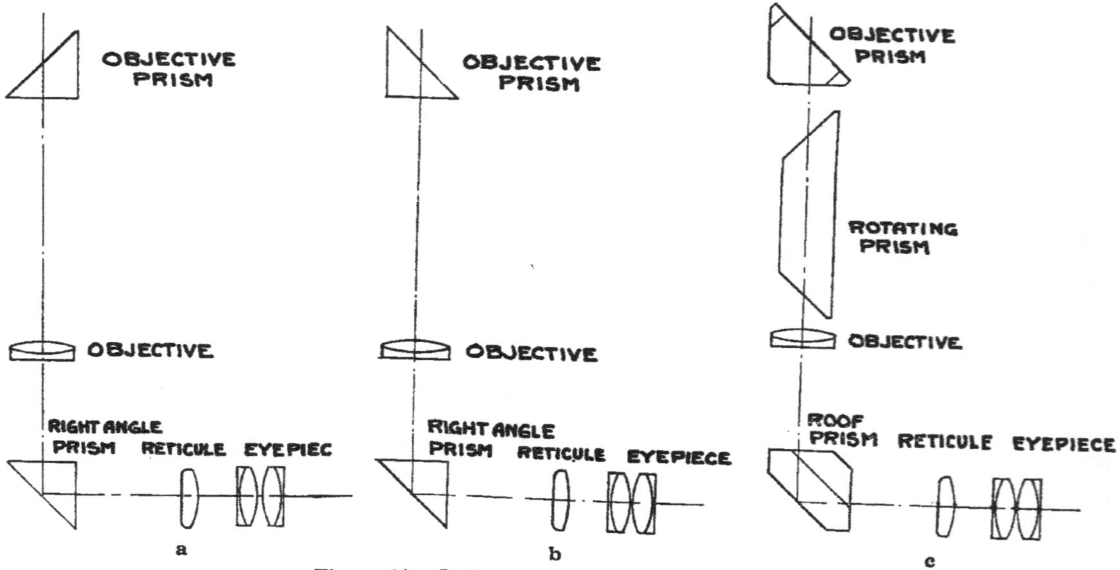

Figure 14. Optical System of Panoramic Sight

THE PANORAMIC SIGHT

81. Will the image be erect in the hypothetical panoramic sight shown in Figure 14 (a), which has two right angle prisms and the objective prism turned for a back sight?
82. Will the image be perverted?
83. If the objective prism is turned around so that observer looks straight ahead, as in Figure 14 (b), will the image be erect?
84. Will it be perverted?
85. What will be the position of the image for an intermediate position of the objective prism?
86. The system actually employed in the panoramic sight is shown in Figure 14 (c). It is to eliminate rotation of the image that the rotating prism is added, which is so geared to the objective prism that it rotates at half the angular velocity of the objective prism. If the two ordinary right angle prisms are retained, will the image now be perverted?
87. How will the substitution of a roof prism for the lower right angle prism remedy this difficulty? Try to picture to yourself the manner in which the rotation of the rotating prism keeps the image erect as the objective prism is turned.

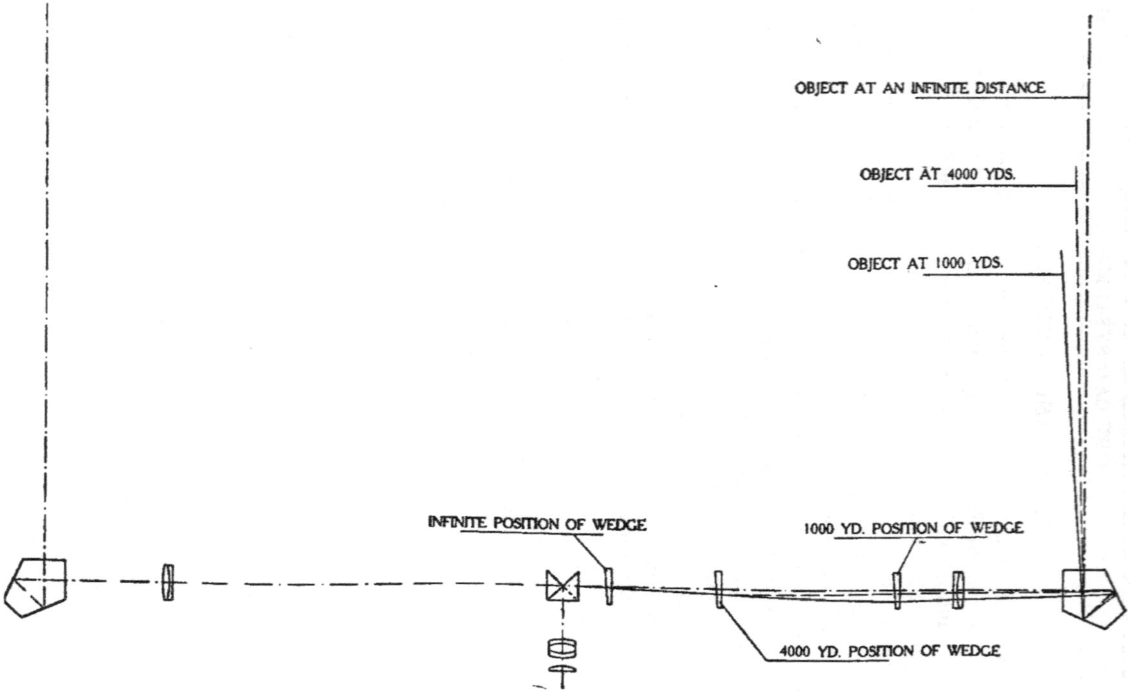

Figure 15. Optical System of the Range Finder.

THE RANGE FINDER

88. Why are penta prisms instead of right angle prisms used at the ends of the base line of the range finder?

89. If an ordinary right angle prism were used instead of the penta prism, through how great an angle would it have to rotate in order to displace the reflected ray 1 second?

90. If the square faces of the prism are 2 x 2 inches, and one of the edges is held stationary, how far would the other edges have to move in order to permit this much rotation?

91. In an instrument of the type which shows one half of the field inverted, do both halves of the range finder have the same number of reflections?

92. With the sharp dividing line employed in the range finder the eye can detect a lack of coincidence between two points subtending an angle of 15 seconds in the apparent field. To what actual angle in the real field does this correspond, if the magnification is fifteen?

93. If the base is 9 feet, and the object is 4,000 yards away, how much nearer must the object be brought in order that its parallax from the two ends of the base line may change 1 second?

94. If the object is 8,000 yards away, how much nearer must the object be brought in order that its parallax may change 1 second?

95. What will be the angular error in the final adjustment if the adjusting lath is one-half millimeter shorter than the base of the range finder and is set up at a distance of 30 meters for adjusting the range finder?

96. In the adjustment of the range finder, it is assumed that the two principal rays, proceeding from the two marks on the lath to the respective ends of the range finder are parallel and make the infinity adjustment. If the range finder has a one-meter base, at what distance from the range finder will these rays actually intersect under conditions of Problem 95?

97. To what linear error will this angular error correspond, if the object is distant 3,000 meters from the range finder?

98. If the adjusting lath is set up 200 meters from the range finder, what will be the angular error in adjustment?

99. To what linear error will this correspond when the object is distant 3,000 meters?

100. The adjusting lath for a one-meter range finder is of the correct length but, instead of being parallel to the range finder, it is rotated about a vertical axis through an angle A, thus producing foreshortening. If the adjusting lath is set up 200 meters from the range finder, how great will the angle A have to be in order to produce an angular error corresponding to a three per cent error at a range of 3,000 yards?

The Engineer School at Camp Humphreys

A REPORT ON METHODS OF
TEACHING ENGINEERING

WAR DEPARTMENT
COMMITTEE ON EDUCATION AND SPECIAL TRAINING
WASHINGTON

4

LETTER OF TRANSMITTAL.

Washington, D. C.,
June 30, 1919.

The Honorable,
The Secretary of War.

My dear Mr. Secretary:

Herewith is submitted a report on the work of The Engineer School at Camp A. A. Humphreys. This has been prepared under the direction of the Advisory Board of the Committee on Education and Special Training in response to an invitation from the Chief of Engineers. The work now being done at this school is pioneer work in higher technical education and it is, therefore, of great significance to the future of engineering education both in the army and in civilian schools. It is recommended that a small edition be printed and distributed to those who may be interested.

Respectfully,

C. R. MANN,
Chairman, Advisory Board.

TABLE OF CONTENTS

THE ENGINEER SCHOOL AT CAMP A. A. HUMPHREYS

Immediately after the Civil War, Congress authorized the maintenance of a permanent battalion of engineers as a part of our regular army. An engineer post under the direct control of the Chief of Engineers was established at Willets Point, New York, where three companies were permanently stationed. In the fall of 1866 plans were discussed for a school of application for officers and men of the Reserve Corps. It was also planned to make the school a special laboratory of the Engineer Corps for investigations and experimental research. The officer instruction at first was carried on through the Essayons Club, a voluntary organization, which encouraged individual reading and preparation of papers on professional subjects for discussion.

The Engineer School at Willets Point developed gradually from the Essayons Club status into a school of application covering the use of astronomical and surveying instruments and important investigations in electricity as applied to the operation of submarine mines. This electrical work at the school resulted in the first practical system of electrically controlled submarine mines adopted by the United States for the defense of its important harbors.

The experimental work associated with the school developed by 1881 to a point that made it seem desirable to detail artillery officers to the school for special training in torpedo work, mine planting, etc. The curriculum gradually expanded to keep pace with the growth of scientific and engineering knowledge. In 1901 the school was moved to Washington Barracks, D. C., where a two year course was given in military, civil, electrical and mechanical engineering. Regular examinations were given and graduation from the course was noted in the army register as an important factor in the officer's military record.

The students admitted to this course were graduates of West Point, who had been assigned to the Corps of Engineers. After graduation they were usually sent for one year's training with troops in the field and then assigned to the school at Washington

Barracks. The method of instruction there was that usually followed in colleges in their seminar courses. Officers were assigned certain courses of reading and attended conferences with their instructors at stated intervals. In connection with this reading course, a considerable amount of practical laboratory work was done in the electrical courses and a quite complete course in practical field astronomy and the use of astronomical instruments in geodetic surveying was given. After completion of the course, the engineer was assigned as apprentice to an older engineer on practical work, and was as a rule not assigned to responsible charge of engineer projects until years after graduation from West Point.

The war interrupted the work of The Engineer School. In its stead a large program of reserve officer training had to be inaugurated. Thousands of men who had had technical training in civilian schools had to be given intensive military training in order to prepare them for commissions and active service with the army. With the signing of the armistice the Chief of Engineers was again confronted with the problem of re-establishing the engineer school to supply the professional training needed to make an engineer officer out of a West Point graduate. This problem was rendered more acute by the fact that the West Point course had been cut off at the end of two years as an emergency measure, and 62 graduates of the two-year course had been commissioned as second lieutenants in the Corps of Engineers with but very slight instruction in the applied sciences. Since the emergency had passed these men had to be trained for service as engineers on the lines of work for which the Corps is responsible in times of peace.

Under these conditions, Major General Wm. M. Black, Chief of Engineers, decided to establish at Camp Humphreys a new school of engineers, with a three-year course of instruction, designed to develop officers who would be able to perform the extraordinarily varied and specialized duties required of the Corps of Engineers. In organizing this new school, General Black desired to apply the lessons of the war-time experience in intensive officer training, as well as the earlier experience of the engineer school, to the problem of training engineers who should be well grounded in the fundamental principles of engineering science and able to tackle any problem which falls in the field of the Engineer Corps and to get results.

In order to accomplish these ends, Colonel V. L. Peterson, who made an exceptional record as commandant of the Officers' Training School at Camp Lee, was assigned as commandant of the new school at Camp Humphreys. A number of engineer officers were detailed to co-operate with Colonel Peterson in planning the curriculum and the course of study. The majority of these officers had no experience with teaching or with schools, but all had demonstrated their ability as practical engineers, and all had clear conceptions of the result that was required.

In the selection of officers to do the actual teaching a similar policy was pursued. Men were selected primarily because of successful practical experience in the various lines of work which they were called upon to teach. Their work as teachers is carefully supervised and frequent conferences are held both among the officer instructors and with the commandant.

The problem and practical application method of instruction, which before the war had been used successfully in some courses in the old engineer school at Washington Barracks, and in some of the best civilian institutions, was adopted as the basic method of treatment. The curriculum was planned to meet the requirements of men who had completed the emergency two-year course at West Point and had been commissioned as second lieutenants in the Corps of Engineers. After only three weeks of planning and preparation, the school was opened on December 7 with 62 lieutenants and 10 captains as students and 20 instructors. Later 18 more captains, who had graduated from the emergency three-year course at West Point and had seen some service with troops, were added to the student body.

The subjects included in the first-year program and the amount of time devoted to each are as follows:

Subject.	Hours per week.	Weeks.	Total Hours.
Mechanical Drawing	10	15	150
Chemistry (class)	12	12	144
Chemistry (laboratory)	5	4	20
English	1	16	16
Engineering Mechanics	12	25	300
Shop Work	8	10	80
Sound and Light	5	9	45
Equitation	2½	20	50

Because of the speed with which the project was executed, the perfection of the details of courses and administration proceeded in parallel with the operation of the school. It has, therefore, been possible to make changes quickly whenever faults appear and to maintain a fine spirit of investigation in the entire enterprise. The whole organization is operating as a well-trained team, determined to achieve a clearly defined result in the best possible way.

Mechanical drawing was intended to supplement the work which the men had already had at West Point. A few simple exercises with lines and circles in pen and ink were given to assure familiarity with instruments. Each student was then required to construct from notes and specifications placed on the blackboard sections of structural steel, crane hooks and chains, nuts, bolts and screws. There was no copy work. Drill in visualizing from drawings was given by working up orthographic projections from isometrics and the reverse. The objects for these exercises were generally rather complicated machine parts. Finally three plates of stereotomy were required. In all of these a student was required to complete the entire drawing, and to make isometrics, templet, patterns, and bevels for the most complicated stones. The final problem of this first stage of the course called for the complete design of a structural steel roof truss, the student being guided in his work by notes published by the school.

About April first the mechanical drawing was replaced by a course of Sound and Light, which treated of the military applications in sound ranging and optics, which have developed so rapidly during the war. This course is a combination of lectures, recitations on assigned readings and laboratory work very similar to the physics laboratory work found in any American college. Each student received a mimeographed direction sheet for each experiment.

The purpose of the work in chemistry was to give the student a clear picture of the chemical industries of the country and a general view of the trend of chemical work. The army engineer does not need the detailed and exact facts which the chemist or chemical engineer must have. He need not be an expert laboratory workman. He should, however, know how, where and why certain products are manufactured, the general outlines of qualitative and quantitative analysis, some organic chemistry, materials to be employed for the manufacture of chemical apparatus, etc. He should also understand the fundamental principles of manufacture relating to the handling of labor and the shipping, storage and costs of raw materials and finished products. He should know that yield figures are of importance only when connected with data on production and costs. He must realize that text-book equations are very incomplete expressions of reactions, indicating results under ideal conditions.

In order to achieve these ends, the course gave less attention to the chemical reactions as such and paid greater attention to commercial methods of production, the actual application of the elements to purposes of warfare, the manner of shipping, storing and preserving. All the elements were discussed in this manner before the general study of compounds began. Lengthy introductions were avoided and the problem presented as it would be in the factory. For example: Given a calcium phosphate rock of known purity; what weight of rock is needed to produce enough phosphorus to fill a given number of smoke grenades of given capacity? In addition to the class discussions, the students spent altogether about 20 hours in the laboratory learning the simple standard reactions and the elementary conceptions of qualitative analysis.

For work in English, the student is required to write reports on subjects with which he is familiar. For example: A trip to the Bureau of Standards was the subject of one report. These reports are criticized and returned for correction and consideration. In order to develop ability in oral English also, the students are divided into sections of 18 each, and each section discusses orally the questions of importance to its members, such as: Should officers be promoted by seniority or by selection? A committee is appointed from the section to investigate the subject under consideration and prepare a report. After the report is presented it is discussed by all members of the section. A vote is taken as to whether the meeting will accept the findings of the report or not.

While the instruction in all departments of the school is in accord with progressive tendencies of current educational practice, the course in Mechanics of Engineering is unique and stands out as a striking contribution to methods of teaching this difficult subject. Full details of this work are given in Part II. It is, therefore, probably sufficient to mention here that the fundamental conception back of the course is to train students to analyze problems by engineering methods and to gain self-confidence so that they do not hesitate to tackle difficult situations. To accomplish these ends the work begins with a real problem—a small highway bridge near the camp—which they examine. In the classroom a simple type of bridge was taken for investigation. A heavy piece of artillery must be drawn across the bridge. Will the bridge stand it? Is it safe?

In trying to answer this question the student soon discovers that he must analyze the problem and then secure tools in the way of fundamental principles and data concerning materials. He is, therefore, glad to be guided in an analysis and to be shown where these tools and this information can be secured. As he proceeds and his collection of principles and information becomes more extensive, he gradually acquires greater confidence in his ability to analyze and meet any ordinary situation. Before the close of the course he is taken through a standard text-book as a review and as a means of organizing into a logical system the material which he has learned by use.

There are many other phases of this mechanics work that are worthy of note. Specifications are issued concerning the manner in which all computations must be performed. No credit is given for an exercise unless the numerical result as well as the method of solution is correct. In the early stages all solutions are made on the basis of fundamental analyses and no derived formulae are used. Later, formulae and standard hand books are freely used in order that the students may secure a proper working knowledge of them.

Early in March it was decided to give the students some shop work in order that they might have a general working knowledge of the essential operations and the time required to complete shop jobs of various kinds, and also that they might become familiar with the properties and limitations of different kinds of machine tools commonly used in working both wood and metal. Since the work was to begin April first, it was decided to follow the outline of exercises that was already in use in one of the well-known engineering schools. In accordance with this plan, the students followed directions and blueprints which would carry them through a series of typical experiences in ordinary shops.

It is planned next year to substitute for these typical exercises productive jobs required for the repair and maintenance of the regular camp machinery and equipment, thereby introducing the problem idea and giving the greater incentive that comes from doing necessary jobs.

In grading the students the standard West Point system is used. Each man receives a numerical grade for every oral or written recitation. The scale used is from zero to three. The arithmetical sum of all grades received during the month determines the standing for the month and the order of merit is that of the sums from the highest to the lowest.

In the mechanics course this system was modified by assigning different weights to different portions of the work. The entire course was divided into five periods, and the marks during the first period were assigned a weight of 1, those during the second period a weight of 2, those during the third period a weight of 3, etc. For the final records only the grades assigned during the last period were used.

The standard army intelligence test was given to all of the

students during the last week of the course. The result indicated that all but two of the students were of grade A, which is in accord with the general record made by the entire Engineer Corps during the war.

The school closed its first year on June 15, and the students were sent to France in charge of Colonel Peterson to make practical studies of the engineer work done over there during the war. The course for the second year will be opened in September, and will consist almost entirely of engineering subjects treated by the method that has proven so successful in the engineering mechanics. From 20 to 30 experienced engineer officers have been working for a number of months preparing in detail the major, minor and development problems which constitute the backbone of the course. Special reports giving details of the curriculum as at present planned and samples of sections of the courses are appended.

In making this study the Committee on Education and Special Training has had the co-operation of the following specialists:

S. L. Conner, Professor of Railroad Engineering, Tufts College, Medford, Massachusetts.

F. H. Evans, Consulting Engineer, Ransom & Randolph Co., Toledo, Ohio.

W. K. Hatt, Professor of Civil Engineering, Purdue University, Lafayette, Indiana.

Wm. A. McCall, Instructor in Educational Psychology, Columbia University, New York.

H. N. Ogden, Professor of Sanitary Engineering, Cornell University, Ithaca, New York.

These experts devoted from one to four weeks to this work, and each of them spent a number of days at Camp Humphreys in a close study of the school.

Part II presents the details of the course in mechanics of engineering as actually given during the past six months. No time for preparation of this course was available before instruction actually began, and therefore this outline is a record of work as it progressed from day to day. The outline must not be taken as a final statement of how the work will be handled because the details are constantly being revised and improved. Part III gives the outlines so far prepared for next year's work in civil engineering. These courses have not yet been given and will doubtless be modified considerably both before beginning the classes and during the progress of the work. Similar series of problems are in preparation for the classes in electrical and mechanical engineering.

During the last week of the course a questionnaire was circulated among the students by Mr. McCall asking their impressions of the relative merit of the Camp Humphreys course and that of West Point, and also their estimate of the work in mechanics. The results indicated that the students saw little difference between the two schools in general, but that the mechanics course at Camp Humphreys was unanimously voted to be more stimulating to interest and speed of learning. These student opinions may be taken for what they are worth, as it is far too early to secure objective evidence that will warrant positive conclusions.

It is also the unanimous opinion of the civilians who studied the work that the methods of instruction developed in the mechanics course are a real contribution to engineering education because they win and hold the students' interest; and that therefore these methods, as it is planned to apply them next year to the engineering courses, are a valuable contribution to the solution of the difficult problem of developing young engineers who are well grounded in fundamental engineering sciences and who know how to tackle a new problem and get results.

PART II

THE COURSE IN MECHANICS OF ENGINEERING

THE COURSE IN MECHANICS OF ENGINEERING

The course in engineering mechanics is designed to develop mastery of the fundamental principles and facility in their use in solving unfamiliar problems. Hence it begins with the analysis of real problems and introduces mathematical and physical principles when the need for them as tools has become apparent. The series of problems are so selected as to require the use of the essential conceptions and principles of mechanics.

The first problems are worked through without a text. Later hand books and tables are used freely. The last five weeks of the course are devoted to a review with a standard text for the purpose of fixing the subject in mind and organizing it in logical form.

In the following pages a portion of the analysis of the first problem is given in detail in order to show the successive steps and methods of procedure. Owing to the fact that instructors soon grasped the main ideas of the method of presentation, subsequent study problems were not outlined so carefully. It was found that the main features were more easily covered in conference at which the instructors made such notes as they needed. Skeleton outlines of these problems are shown on the chart, in which the organization of the course and the analysis of the separate problems are indicated graphically.

These outlines are significant because they indicate the novel method of organization. The main feature lies in the fact that the unifying center of each problem is a material structure, such as a bridge, a water tower or a dam, and that around this center are grouped numerous mechanics concepts and principles. In ordinary methods of mechanics instruction the subject-matter is organized about a mechanics principle as a center, and then some special details of an engineering structure are cited as examples of the application of the principle in question. The outline of the first few weeks instruction in the mechanics work will be found on page 21.

Analysis of the outline shows that the method of procedure was as follows:

(a) A problem was described and presented.

(b) The instructor stimulated the students to go as far in solving the problems as they could without any other aid than their own knowledge.

(c) The students were helped in overcoming recognized obstacles when they had spent as much time as was profitable under the circumstances in attempting to overcome them. A student may be helped over an obstacle in one of two ways. One way is for the instructor to give a demonstration or even a hint that will lead the student to apply principles used in previous solutions. The other way is to give a practical method without proof for the student to use. In a later problem the theorem of three moments was presented in this way. Students desiring the proof found it much easier after the theorem had become familiar through being used in solving actual problems.

(d) When a new principle has been brought to the student's attention during the solution of an engineering problem, an effort is made to fix the principle in the student's mind by presenting other problems, not of the same type, but involving new situations in which he must use the same principles. In this way idea memory is obtained rather than mechanical memory.

All work which the student turns in must comply with the specifications given on page 33.

A text, Poorman's Mechanics, was introduced near the last of the course. This tied together what had been covered previously and provided a ready reference in addition to the student's notebook.

OUTLINES OF CLASS WORK IN MECHANICS OF ENGINEERING

The first meeting of the Engineering Mechanics Sections was held on Saturday morning, December 7, 1918, at the bridge over Accotink Creek, near the village of Accotink.

The attention of each member of the class was called to the following information concerning the bridge, each part or feature being explained:

Highway bridge.
Pony truss bridge.
Parallel chord bridge.
Trusses and truss members.
Chords, top (or upper) and bottom (or lower).
Web systems and web members.
End posts, posts, columns, struts.
Hangers.
Main diagonals, ties.
Counters.
End shoes.
End bearing plates.
Anchor bolts.
Floor planks; also longitudinal planks for reinforcing floor.
Stringers.
Floor beams.
Lower lateral system of wind bracing.
Span of bridge, center to center end bearings.
Depth of bridge, center to center of chords.
Width of bridge, center to center of trusses.
Panels, panel points (upper and lower).
Guard rails and hand rails.
Loop rods.
Turnbuckles.
Upset rods.
Tie plates or batten plates.

Lattice bars.
Pin joints. Pin plates.
Riveting. Bolts for erection of bridge.
Channels, angles, "I" beams, round bars, square bars.
Connections of one part of bridge to another.
Splice in top chord.

The following defects in the bridge or departures from good practice were noted:

Rusted plates.
Bent floor stringer.
Bent rod in lower chord.
Lower later diagonals not in plane of lower chord.
Tie plates too far apart.

The class was then divided into four sections of 18 men each and one section at a time was taken onto the bridge and put thru the following quiz:

Why are some members rod and some members built up out of shapes?
What is best form for a tension member?
What is best form for a compression member?
Why is top chord so wide?
Why are lattice bars and tie plates used?
Why are counters put in? When do they act?
Why are the counters "Upset" at the turnbuckle?
How does the "Load" get from its point of application to the ground? Trace its path from member to member.
How does force go from one member to another when both are connected to a pin?
How is floor fastened on?
How may a weak floor be reinforced?

Attention was called to the great importance of making all connections between parts of the bridge thoroughly safe.

Special effort was made to have each member of the class clearly understand the action of a "Pin Joint" and how stresses are transferred from one member to another through the pin and how the pin is subjected to shearing and bending forces.

THE FIRST STUDY PROBLEM

The second meeting of the class in Engineering Mechanics was held on Monday morning, December 9, 1918. Four sections of nine men each met from 8:00 to 9:30 and four sections from 10:15 to 11:45.

A King Post-Pony Truss-Highway Bridge, was taken up as the first engineering problem.

The following notes were given as suggestions to the instructors, with the understanding that the details of the work should be changed to suit conditions, always keeping the main object in view, that is, the development and training of the student.

Each instructor was expected to supplement the work outlined in the notes by bringing in additional illustrations and explanations derived from his own experience. In other words, the individuality of the instructor was to be used to the fullest extent possible in order to make him an effective leader or coach for his men.

In the first problem the following mechanics concepts are introduced:

Weight of Timber and Steel.
Forces.
"Free Body" idea.
Things necessary to know in order to know a force.
Force represented by an arrow.
Definition of "Two Force Piece."
Transmission of the effect of forces by means of triangular frames.
Direct Tension and compression.
Internal Stress holds in equilibrium the External Loads.
Unit Stress, Ultimate Stress, Proportional Limit.
Factor of Safety.
Upset rods—reason for, and design.
Investigation and design of steel bars in tension.
Laws of Equilibrium for forces in a plane.
Summation of components of forces in any direction equal to zero.

Graphic determination of stress, three forces at a point, in a plane.

Analytic determination of same. Resolution and composition of forces.

Investigation and design of long columns in timber by straight line formula.

Physical ideas of Bending and Shear.

This problem was used as a study problem for four days, December 9, 10, 11, and 12, 1918.

TEACHER'S NOTES ON THE FIRST STUDY PROBLEM

Before every session of the class a conference of instructors was held for the purpose of drawing up a series of questions to indicate the order in which the various ideas might best be presented. Under each question notes were made to suggest to the instructors subject matter that should be mentioned. The following development questions and notes, used in the first lessons are appended as a sample of the method.

Similar questions and notes were made for the analysis of the other parts of the bridge. After a few class sessions the instructors became used to the method. Then no formal notes on the conferences were needed.

Span 24′ C-C end bearings (2 panels at 12′).

Depth, 6′; width, 16″ C. C. Trusses. Concrete abutments on rock foundation.

Bridge is over gorge 100′ deep, with very steep, rocky sides.

Upper Chord (End Posts), 4″x6″ timber.

Lower Chord, ¾″ round bar, not upset.

Hanger, 1″ round bar, not upset.

Floor beam, 2′-8″x14″ timber.

Stringers, 4″x12″ timbers, 2′0″ on center.

Floor Plank, 4″x10″ timber.

Guard rail, 4″x6″ timber.

Sill on abutment, 6″x6″ timber.

Question: Military necessity requires that a 12-inch howitzer (fully loaded for traveling) be taken across this bridge.

Question: Is the bridge safe?

NOTE—Suggest that students make notes of ideas that come to them while thinking about this.

NOTE—Encourage them to try—to start something.

Question: In how many ways is it possible for bridge to fail? List them.

NOTE—Follow load from point of application until force gets into ground. (Floor plank, stringers, floor beam, hanger, upper chord, shoe, bearing plates, abutment, and all connections between these parts, also lower chord rods).

Question: How may the floor planks fail?

Question: Make sketch showing worst position of live load on planks.

NOTE—The analysis of bending stress in the planks can be postponed temporarily.

Question: If planks were too light, how could they be quickly strengthened?

Question: How may the stringers fail?

NOTE—(Carry this through in same manner as for floor planks).

Question: How may floor beam fail?

NOTE—(Carry this through in same manner as for planks and stringers).

Question: What shall we do in order to find out whether the hanger is safe?

Question: How much load at lower end of hanger rod?

NOTE—(Dead load: Weight of planks, stringers, floor beam and guard rail, which properly goes to hanger).

Question: How get weight of timber? How figure F. B. M.?

Question: (If live load has been forgotten)—How about effect of live load?

Question: What position of howitzer will give greatest load on hanger?

Question: What is total load on lower end of hanger? What load on upper end?

Question: How about the effect of Impact?

Question: Take hanger out as "Free Body" and show forces acting on it.

NOTE—A "Free Body" sketch is one which shows some particular body by itself with all other bodies previously connected with it taken away, the effect of the removed bodies being replaced by the proper forces (represented by arrows).

NOTE—Never take away any part of a body without putting in a force to represent it.
Never put in a force where nothing has been taken away.

NOTE—A force is fully defined when the following is known:

(a) point of application.
(b) direction in space.
(c) sense, or direction along action line.
(d) amount, or magnitude.

A force may be represented by an arrow which denotes these four things.

Question: What force is acting at upper end of hanger?

NOTE—If weight of bar itself is included by any students, have them figure its weight and compare with the load on hanger. A bar of steel one square inch in cross section and one foot long, weighs 3.4 pounds. (In this case, weight of bar may be neglected, but the student should arrive at this conclusion himself).

NOTE—The rod in the bridge doesn't move (is in equilibrium). The free body in sketch represents the rod in the bridge, therefore the free body does not move (is in equilibrium, etc.).

NOTE—Conditions for equilibrium of two forces are:

(a) must act in same line.
(b) must be opposite in sense; and
(c) must be equal in amount (magnitude).

NOTE—A rigid body which is in equilibrium under the action of two forces only is technically called a "Two Force Piece."

Question: How much force is acting internally in the bar?

NOTE—Have two men pull on a cord and ask whether tension in cord is different from pulls at each end. Substitute spring balances.

(Take out parts of the hanger as free bodies. Wherever the section is made the following will be true).

NOTE—Internal stress in bar holds in equilibrium the external load on bar.

NOTE—The internal stress in a "Two Force Piece" equals the External load at EITHER end of the piece. (If weight of piece is neglected).

The hanger is a case of direct tension.

Tension stress is uniformly distributed over area of cross section, therefore:

(Total Stress) equals (Unit Stress) times (area of Cross Section).

Question: Is the hanger safe? What is your conclusion?

NOTE—A kip equals 1,000 pounds; abbreviation——"k."

NOTE—Compare unit stress with ultimate tensile stress (60 kips per square inch).

NOTE—Unit stress should be compared with proportional limit rather than ultimate stress. When proportional limit is reached the danger point is reached. Proportional limit of this steel is about one-half its ultimate strength or (30 kips per square inch).

Question: What is proportional limit? Elastic limit?

NOTE—The recommended use of proportional limit instead of elastic limit to mark the end of straight line variation in stress and deformation.

Question: (Discuss Factor of Safety).

Question: (Discuss—upset rods, and area at root of thread).

NOTE—Allowed safe stress equals 16 kips per square inch, giving a factor of safety of about 3.75.

NOTE—The hanger stress is higher than allowed in good practice, but below the proportional limit, and could be called safe for this particular emergency.

Question: What size should hanger be to conform with good practice?

Question: Is hanger safe if howitzer gets off the middle of roadway and runs near side of roadway?

THE SECOND STUDY PROBLEM.

Report on the Safety of a Timber Tower Supporting a Water Tank.

The principles involved in this problem being largely the same as in the first one, no outline was prepared for the instructor's use. The problem was discussed by the instructors in conference on December 12, particular attention being given to the new ideas which the problem brought in.

This problem involves the following mechanics concepts (those introduced for the first time are marked with a *) :—

* Moments.
* For Equilibrium, moments of all forces equals zero.
Weight Timber and Steel.
* Weight of Water.
* Wind pressure on Flat and Cylindrical Surfaces.
Application of the Laws of Equilibrium.
* Increased reactions due to wind; also uplift due to wind.
* Pressure on Soil and Safe Bearing Power of Soil.
* Investigation of safety of timber in side bearing.
* Investigation of safety of concrete in bearing.
* Determination of stresses in truss.
Investigation of timber columns.
Physical ideas of bending and shear.

This problem was used as a study problem on December 13 and 14, 1918.

THE THIRD STUDY PROBLEM

Report on the Safety of a Concrete Dam.

The design of this dam was submitted in connection with a proposed water supply system for a city located in the southern part of the United States.

This problem involves the following mechanics concepts (those introduced for the first time are marked with a *) :—

Weights of concrete and water.

* Hydrostatic Pressure.
* Hydrostatic Prism.
* Friction of Rest.

Application of Laws of Equilibrium.

Factor of Safety.

* The insertion, in a given system, of two equal and opposite forces in the same line does not change the previous condition of equilibrium.
* Couples in the same plane are equal if their moments are equal and in same direction.

Direct Stress (or Uniformly Distributed Pressure).

* Bending Stress (or Increased and Decreased Pressure due to tendency to overturn).
* Combined Stress (or Resultant of Direct Stress and Bending Stress. Actual distribution of Pressure on base of dam).

Safe Bearing Power of Soil.

This was used as a study problem on December 20 and 21, 1918.

THE FOURTH STUDY PROBLEM

(A) Investigation of the Dam in the previous problem as to safety at any section between the top and bottom.

This problem furnishes drill in all the features brought out in the previous problem.

The special purpose of the problem is to give a clear physical idea of Shear and Bending Moment, which are here introduced.

(B) 1. This problem also introduces the following mechanics concepts:

* Diagram showing intensities of water pressure at various depths. (Load Diagram).
* Diagram showing tendency of the dam to slide off at various horizontal sections. (Shear Diagram).
* Diagram showing tendency to break off at various horizontal sections. (Bending Moment Diagram).
* Idea that shear is rate of change of moment.
* The three Laws of Derived Curves.

2. The profile for an Electric Transmission line used as a means of introducing the idea of "Rate Curves," and as drill in first and second laws of derived curves.

3. Profile of vertical curve, for a railroad track, and its rate curve used to emphasize the idea that an inclined straight line in one diagram creates a curve in the next higher diagram.

THE FIFTH STUDY PROBLEM

Investigate the Safety of a Timber Cofferdam to be Used in Construction of Pier in a River.

In this problem the following concepts of mathematics and mechanics are involved (those introduced for the first time are marked with a *) :—

Hydrostatic Pressure.
Load Diagram.
Shear Diagram.
Moment Diagram.
First Law of Derived Curves.
Second Law of Derived Curves.
Third Law of Derived Curves.
Center of Gravity of Triangle.
Moments.
Reactions.
Parabola (area).
Parabolic Spandrel (area).
Quadratic Equations (solution of).
* Cubic Equation (solution of).
* Inflection Point of Bending Moment.
Couples.
* Investigation Fiber Stresses in Timber Beam.
* Safe Fiber Stresses in Timber Beam.
Investigation Timber Columns.
Straight Line Formula for Timber Columns.
Investigation of Bearing on side of Timber.
Safe bearing on side of Timber.
* Investigation of Longitudinal Shear in Timber Beam.

SPECIFICATIONS FOR MECHANICS WORK

(Supplied to all students.)

1. Unless otherwise specified, use only Standard Mechanics Paper; size, 8½"x11"; color, yellow or orange; punched for Standard I. P. 3-ring Binder.
2. Holes for Ring Binder shall be at left of sheet.
3. Margins shall be ruled as shown on Sample Sheet. Problem Number, Name, Section Number and date shall be placed as shown. (This should be the first work placed on each sheet).
4. Begin each problem on a new sheet. Not more than one problem shall be put on a sheet unless otherwise instructed.
5. Work shall be arranged systematically and clearly on each sheet. (By doing this, errors which grow out of confusion will be avoided).
6. Explanatory headings shall be used throughout to indicate steps taken in the work. Marginal index shall be used to identify results obtained in the various steps. (This leads to accuracy and efficiency and increased speed. It also saves time and effort when checking or referring back to the computations).
7. Rule a space at the left of the sheet (as shown on sample), for the segregation of the so-called "scratch-paper" work. Vary the width of this space to suit the nature of the work. (This work is of great importance and must be done with the same clearness as the other work or else accuracy will be sacrificed. Proper care is good insurance).
8. Rule horizontal lines entirely across the sheet to separate parts of a problem (or where there is a break in the work). Do not rule vertical division lines.
9. Use no short cuts. Perform one operation at a time and make a complete record of it.

(Neat and careful work leads to neat and careful thinking and vice versa).

10. All work shall be done with a pencil which gives a clear and black mark. Enough pressure must be put on the pencil to make the marks black. Pencil points must be kept sharp so as to give clear marks. (Pencils which are too hard or too soft will not give satisfactory results).

11. All sketches shall be made with triangles or straight edges unless otherwise specified. Outlines must be clear and black. Strong contrast between real and imaginary lines.

12. Dimension figures must always have the right of way and must be kept clear of all lines.

13. Decimals must have a distinct point. (The decimal point is of such great importance, and errors from its omission are so costly, that it is a good plan to exaggerate it).

14. Fractions shall be written with a horizontal line and not with an inclined line.

15. Statements of "Proportion" shall be shown in fractional form with the sign of equality and not with proportion signs. The unknown shall be placed in the numerator on the left side of the equation.

16. Make large sketches. (The sketch is an aid to clear thinking, not merely a picture of what has already been done. A "working sketch" is only relatively to scale. Certain parts should be exaggerated in order to make the work convenient and easily understood).

17. Use only Standard Engineering Lettering and large figures for all problem work.

18. Stresses in framed structures shall be noted on the proper members of a "Space Diagram" (or "Stress Diagram"). The Space Diagram must be large enough to show this information clearly. Compression Stresses shall be marked C; Tension Stresses T.

19. Answers to problems and the results of various steps of a problem shall be made prominent by the device shown on sample sheet, except that no such device shall be used for

results noted on diagrams (such as stress or curve diagrams).

20. Units shall be clearly indicated for all given data and for all results.

21. Accuracy is absolutely essential in engineering work. (Precision may not be necessary). The engineer who cannot be depended upon for results is worthless. The student must learn the cost of errors. Incorrect results; no credit.

22. Use check methods whenever possible and surround all work with as many safeguards as can be used efficiently.

PART III

COURSES IN STRUCTURAL ENGINEERING

COURSES IN STRUCTURAL ENGINEERING

The courses in Structural Engineering are under preparation, and will be given for the first time beginning September, 1919.

Classified as Structural Engineering are the following:

Subject.	Number of Periods.	Length of Periods.	Total Hours.
Materials of Construction	23	3 hours	69
Roads and Pavements	26	2 hours	52
Foundations	26	2 hours	52
Wharves and Piers	15	2 hours	30
Plain Masonry	18	2 hours	36
Reinforced Concrete	12	2 hours	24
Roofs and Bridges	25	2 hours	50
Building Construction	24	2 hours	48

The subjects are pursued intensively, in the sequence as listed. A morning period is 1 hour 50 minutes, and an afternoon period 3 hours in length. A morning subject occurs three days in the week, and alternates with another, and an afternoon subject alternates with the same other during 5 days, 3 periods one week and 2 the next.

The time allotted is tentative. The school year is 11 months. These courses are preceded by training in Mathematics and Mechanics.

These courses are being prepared by engineers of extended practical experience but of no teaching experience with army reserve commissions, on active duty at Camp Humphreys. They are working under a general plan as fixed by Major General Black.

The preparation of each course includes:

A frame work of major and minor problems, through which the student approaches the text.

Writing of school texts

Selection of reference texts

The school texts contain a discussion of the elements that control the design or method of construction, and are unusually valuable in that they bring to the student's attention many important practical considerations that are absent from the ordinary text.

The distinctive feature of each course is one or more major problems, constituting a framework upon which the student hangs the subject matter of the text. The major problem is generally a complete structure that the student is to design, or to specify its method of construction. A complete description of the situation is given. In his progress, he is required to make decisions, and is thrown back to the text for the facts, or theories, or principles that control the decision. For this purpose, the major problem is further analysed into a number of minor problems that are parts of the whole and serve to keep the student's interest alive, and to guide him to a reasonable degree in the sequence of the necessary decisions in the face of a real situation. The latter are brought out by a series of development questions.

This is the job method of teaching and is applied to the engineering courses ordinarily listed for the senior year in engineering schools.

These courses offer a hopeful solution of the difficulty of teaching such courses as Foundations, Wharves and Piers, to the student in the civilian engineering school. It is the general experience that such courses have not the same value as courses in Applied Mechanics, when considered as educational tools; it is difficult to get reaction from the student.

Generally the civilian student informs himself by reading texts, and the subject is discussed, and lectured upon by the professor in class on the basis of the text. A continuous reaction from the student is not obtained. The matter is descriptive. The actual process that the engineer goes through in deciding between conflicting conditions of the problem, or choosing controlling elements, are not generally evident from the text. Nor has the professor as a rule such live and real problems as are available in the course in Structural Engineering now being prepared for The Engineer School at Camp Humphreys.

Instead of reading and discussing a text for classified knowledge upon which he is examined, the army engineer student goes back to the text for the material for a definite decision, for facts, or for a tool to accomplish a purpose.

In civilian institutions, also, courses such as Bridges and Buildings require from the student an extended performance in the drawing of the details. In the courses in The Engineer School at Camp Humphreys the designs are sketched, but not drawn to complete detail. These courses thus take the middle ground between reading courses and detailed design courses. They represent the work of the engineer rather than that of the drafting office.

When looked at from the standpoint of the courses as ordinarily taught in civilian schools, it might appear that too little time was allotted to the subjects; and that a superficiality, which has resulted in the past from a too general reading of a treatise, might again result. But these courses should correct the superficiality. In the first place, the time-consuming detailed drawings of structures by the student are omitted, as has been said, and sketches are used. In the second place, the student is led into the background of the school text for decisions. And this school text is a short summary of the controlling elements rather than a detailed treatment. An instructor should be able to direct the use of the fuller reference treatise when necessary.

It should be said also that under the conditions of the student's life at The Engineer School, he is less subjected to distractions such as those that diminish so greatly the effectiveness of student time in civilian colleges. The physical conditions of the engineer student will be favorable. He is also a very carefully selected man. His power of attainment will correspond to that of our civilian students in the training camps.

At first glance, the decisions expected of students seem to be beyond their capacity, and to demand a judgment that could only be obtained from experience. Professors who have tried similar courses are surprised at the extent to which many students reach wise decisions after a study of the situation and the text. The instructor is furnished with a guide to the correct decisions, and in turn can guide the student. The process of approach to a decision, the knowledge of controlling elements, are learned, even if the decision is incorrect.

These courses must also be viewed in the light of the product that it is desired to produce; namely the engineer officer. It is the usual expectation that the graduates of a civilian school of Civil Engineering will immediately and fairly continuously deal with the detail of a subject, as in surveying operations, or in a designing, or detailing office of a bridge or building construction company. He must learn the detailed processes.

The army engineer, however, has a much wider range of duty. He may have to select the equipment for a power plant, or deal with a difficult foundation, and may do this years after graduation, and only once. He should be trained to handle the problem from the broad standpoint, without the "tricks of the trade," or skill in details. That is to say, he must be trained as an engineer, rather than as a draftsman.

It would seem that such a broad understanding of the various fields of construction, and an ability to think through these problems would be developed by these courses, with an avoidance of the superficiality and lack of interest and reaction that comes from mere reading of treatises.

In brief, these courses aim to accomplish the following:

The student must think out a decision in the face of a real situation.

The subject matter is presented to bring out the fundamental principle, and to force the student to analyze the problem to its controlling elements.

He is continually confronted with a problem, and goes from this to the text for a decision only after exhausting the possibilities of his own experience and judgment.

The practice and the types of structure are presented. No doubt when the courses are operated, adjustments of subject matter and coordination of the courses will be necessary. The relation of the laboratory for testing materials to the courses should be studied. The general question of the student's critical study of data upon which design rules are based, and the measurement by the student of the forces of nature, and the subject of inspection trips to engineering works, will have to be considered.

For the benefit of civilian instructors, in engineering schools, extracts from text and problems of some of these courses are given on page —, under the following headings:

A. The topical outline of the course in Masonry Structures followed by:

I. The statement of the first major problem, the Black Warrier Dam, with the analysis of the first of the minor problems into development questions. See page 47.

II. Extracts from instructor's notes on the first minor problem. See page 50.

III. The statements of the subsequent minor problems. See page 53.

B. The outline of the Section on Timber from the course on Materials, followed by:

I. The analysis of the major problem into minor problems and development questions. See page 68.

II. Extracts from instructor's notes on the first of these problems. See page 73.

The text is a very excellent and well-balanced discussion of quality and use of wood. The instructions to the teacher will aid in a practical design and guide the student to a wise decision.

The discussion of flooring, sash, framing, etc., brings in the qualities of minor species.

Cast iron and steel enter in columns, wrought iron in hangers, wood for tanks, considered in sprinkler system. Shrinkage allowed for.

The text of the courses and the references take the student into the consideration of the elements necessary for a decision. There is supplied to the teacher a discussion and reasons for the correct decision upon each element of the problem. By these an instructor other than the experienced constructor who has prepared the course can satisfactorily administer the course. The student's problem is well analyzed into a framework of subsidiary questions so that he will not be too much at sea. A record of these, with the instructions to the teacher, follows.

This course is marked by a well-matured plan for bringing the student to a decision upon the controlling elements of the problems, and sending him to the background for laws and data upon which decisions must be based.

The time allotted to this course consists of 12 hours in class, 6 hours in laboratory, and 6 hours of night study. The instruction extends over a period of one-half month, with three periods each week of three hours each.

A. TOPICAL OUTLINE OF THE COURSE IN PLAIN MASONRY STRUCTURES.

I. DAMS.

1. Historical (brief).
2. Classification of dams.
3. Types considered.
4. Design of dams.
5. Preliminary investigations.
6. Stream diversion.
7. Foundation, hard.
8. Construction of rubble masonry dams. (brief)
9. Procurement of materials.
10. Construction plant and equipment.
11. Cyclopean masonry.
12. Delivery and deposition of materials.
13. Expansion joints.
14. Drainage of foundation and dam.
15. Overflow dam.
16. Dams on pervious foundations.
17. Construction plant.
18. Present tendencies in construction.
19. Estimates and costs.

II. RETAINING WALLS.

1. Kinds of retaining wall.
2. Method of treatment.
3. Equivalent fluid pressure.
4. Surcharge.
5. Stability of retaining wall.
6. Foundations.
7. Design of plain concrete walls.
8. Foundation for walls.

III. ABUTMENTS.

1. Kinds of abutments.
2. Abutments on waterways.
3. Design of abutments.

IV. BRIDGE PIERS.

1. Requirements in placing bridges and piers.
2. Definitions.
3. Functions of bridge piers.
4. Forms and dimensions of piers.
5. Construction of piers.
6. Hollow piers.
7. Stability.
8. Failures.

I. STATEMENT OF THE MAJOR PROBLEM

The Black Warrior Dam No. 17 is located on the river of the same name, 343 miles above Mobile, Alabama. It is the seventeenth lock on canalization of the Tombigbee, Warrior, and Black Warrior river system which is to serve the coal fields of Alabama. The purpose of Dam No. 17 is to provide the slack water navigation, 6′ minimum draft up the Locust and Mulberry forks. The dam is mounted by a flight up two locks, each 52′x258′, 31½′ lift.

The principal part of the course on masonry structures will consist of the design and study of the construction features of the 65′ overflow dam such as was built here. The conditions will be assumed as they were actually found and the study carried forward to arrive at the structure as it now stands. As the elements are considered different conditions will be assumed in order to bring out various principles of design and construction.

It is decided to improve the river so as to raise the pool level No. 17 to 63′ above the level of pool No. 16. A careful study of the river indicates that the best site for the proposed dam is at Squaw Shoals.

The river at this point has an estimated maximum and minimum flow of 140,000 and 150 cu. ft. per sec., the floods occurring usually during the winter and spring and at rare intervals during the summer and fall. The variations are often very rapid, a rise of 2′ per hour being common. Smaller freshets occur usually several times each dry season. The low season is from June to December. Surveys referenced to lock No. 16 show high-water line to be at elevation 192. Lock No. 16 being unfinished does not obstruct the natural flow of the river.

At Squaw Shoals the river flows between high steep banks. A superficial examination of the site apparently shows a hard sandstone surface across the river bed rising to a bank on each side, at elevation 200. The rock is suitable for concrete material. The elevation of the bottom of the river is approximately 177 at the dam site. The hard sandstone ledge outcrop can be clearly traced along both banks covered with an overburden of sand, clay, disintegrated shale and rock. The banks and surrounding hills are covered with a growth of medium sized timber.

MINOR PROBLEM No. I—INVESTIGATION OF THE SITE

In making the river study prior to locating the dam, it is, of course, necessary to acquire a knowledge of the flow, establish bench marks and take topography. This study, in connection with the considerations governing the general navigation project, has decided the desirability of placing the dam at the indicated location. It remains to determine the practicability of building the dam. It is necessary to investigate the foundation. This will determine what type of dam, or if any dam, is suitable to the location.

1. Since it has been decided that Squaw Shoals is the best location, and the terrain, as it can be seen, indicates approximately the most desirable point, what facts about the proposed location must first be determined?

2. Describe how you would make these investigations.

3. The river bed is about 1200′ wide at this point. At low stage the water flows down in broken riffles as is seen in the drawing (furnished the student.) The topography of the site is shown on the map (Dr. E 2—24.)

Draw a sketch showing how and in what sequence you would proceed with the complete investigation. Assume that the topography has been taken, that there is one bench mark and a base line established. Neglect the detail methods of the survey but show the lines necessary. Number consecutively the points where you take data.

a. How would you get the data in the river bed?

b. How deep would you go?

4. The strata is uniform and level. How would it affect your investigation if many seams were visible and if the strata were inclined and broken?

5. You have observed that the rock on both banks is of the same kind, uniform and apparently the same as the river bed. How does this effect your judgment as to the desirability of the dam site?

a. Suppose the rock on the two banks is different. What does it indicate and what precautions should be taken in the investigation?

b. How may it affects the design?

6. The borings taken and the data obtained are shown on Dr. R5, Sheet 2. The conditions appear to be uniform. Are the borings adequate? State reasons briefly.

7. Additional borings taken are shown on Dr. D2-2. (All necessary drawings are furnished the students.)

a. Are these borings sufficient?

b. Are they spaced close enough?

8. The investigation is concerned with determining the adequacy and second the water-tightness of the foundation. Core borings will give an accurate cross-section and answer fully the first. Much information as to the second condition can be obtained from the borings, by watching the course and run of the drill, particularly if an experienced drill runner is employed and a careful log is kept. The best information can be obtained by testing the foundations for tightness.

Adequate pumping capacity and an air compressor capable of supplying 400 cu. ft. per minute at 100 lbs. are available. The water is running rapidly down over the shoals.

a. How would you proceed to test this foundation?

b. If the work were in still water how would you make the test?

II. INSTRUCTORS' NOTES ON MINOR PROBLEM No. I

INVESTIGATION OF THE SITE

The student will have had instruction on the subject of test pits, wash and core borings, and other means of investigation, but a general discussion of this subject, particularly the difficulties to be encountered in sand and boulder beds, will be carried on during the consideration of this section.

1. The investigation of the site should be made in the following sequence:

a. Whether or not rock exists, over the whole site and if it is suitable for the foundation of a gravity masonry dam.

b. If the rock is tight, or broken and fissured.

c. The depth, character and amount of overburden of unsatisfactory material.

d. Source of supply, location, quantity and quality of construction material such as sand, gravel, stone, etc.

Investigation of a and b will determine whether a solid masonry dam is possible. a, c, and d will obviously have a direct and possible determining influence on the cost and practicability of the structure.

2. It will first be necessary to decide on some comprehensive plan of obtaining the data, the extent of the area to be investigated and lay out base lines. Study should be most intensive under the probable site of the structure, but should also include a considerable area above and below the site, the abutments, and possibly the walls of the valley which will be submerged. Physical conditions such as outcrop of rock, topography, etc., may make any investigation of the banks except immediately adjacent to the site, unnecessary.

Seattle, Washington, built a great and expensive masonry dam on the Cedar River. The site of the dam was carefully investigated but through insufficient data on one bank of the reservoir and misinterpretation of the observations made, the whole project was an absolute failure. The water flowed through the banks of the reservoir into an adjacent valley some miles distant.

Much information as to the banks, abutments and areas above low water can be obtained by test pits, drive rods.

Wash borings are useful, relatively cheap and convenient but are of value only in determining if rock exists and the character of the overburden and depths. For important structures to be founded on rock neither test pits nor wash borings can be depended on as they give no information regarding what lies below the rock surface. If by observation, wash borings, or other means, it has been determined that rock exists over the whole site and the structure is of importance, core borings will be necessary.

Wash and core borings should not be made in sequence and close proximity across the site. They should first be taken at widely scattered points, as a few may give the information desired. This is particularly true if the results show absence of rock. The remaining borings should be taken in such a manner that each shall give the maximum information as to the whole site. It will be unnecessary to put all holes down to the same depth. A few holes widely spaced can be run down to greater depths and give reliable information as to the under strata. The results obtained should give an accurate cross-section of the material existing in the foundation. In general, the higher the dam the deeper the investigation should be carried.

3. Some base line, approximately parallel to the river bank and in such a position as to view the whole site, if possible should be laid out. From this two parallel lines, one at the selected crest of the dam and one near the toe and coinciding in direction with the dam should be turned off. On these a line of drill holes will be put down. If the data does not show uniform conditions probably one or more additional lines will be needed under the dam.

A line of holes at greater intervals, and parallel to the dam, should be put down at about 100 ft. up stream, say at 100 ft. apart across the river, and a few holes drilled further up stream. Investigation down stream from the dam is fully as important, due to the abrasive action of over-fall, but need not go so deep. Durability, not water-tightness, is here of importance.

4. The holes should be first spaced at wide intervals, a few all

over the site. The spacing of intermediate holes is governed by the results shown. Consistent, uniform conditions require fewer holes than when there are irregular, broken, or inclined strata. Inclined strata may indicate a more impervious foundation than a horizontal one, depending on the character of the seams between layers.

5. Rock uniform on both banks of the stream is a very good indication, especially if the same as the stream bed, as no serious faults are liable to exist. Different rock on the two sides, or the presence of different kinds on both banks, may indicate planes of weakness at their contact and the investigation must be more careful. It should be emphasized that the character and extent of the study is determined by the conditions encountered. A broken or fissured foundation will probably require a much deeper and more carefully constructed cut-off, as will be described later in the course.

6. The borings are in no wise adequate. None of them go deep enough to show what is below the solid rock. No data is obtained as to the condition of the river bed above the dam nor enough below where the bed will be subjected to severe over-fall abrasion. The Austin, Texas, failure was due to insufficient protection below the dam.

7. The very uniform conditions found make the wide spacing and the depth of the borings entirely sufficient. The borings are deep enough to give all the necessary data.

8. The foundation in the case of the Black Warrior must be tested with water, as compressed air cannot be depended upon to show in the broken running water. The amount of water which the hole will take and the pressure required, is a good indication of the tightness of the rock. By testing the holes at different elevations, information as to the location of the faults and leaks can be obtained. In still water, tests with compressed air are often very valuable in showing the location and extent of foundation leaks. This can sometimes be well established by the use of aniline dye solutions.

III. STATEMENTS OF THE SUBSEQUENT MINOR PROBLEMS.

MINOR PROBLEM NO. II—UNWATERING THE SITE

The adoption of a proper construction program is a problem requiring the most careful study and mature engineering judgment. The problem is peculiar to each project. It must be worked out as the design proceeds and before even the costs are figured, if the estimate is to be more than an intelligent guess. Deciding on a correct construction program is not only an engineering problem but often decides the financial success or failure of the project. In general, unwatering the site represents the most complex phase of dam construction. Some problems are comparatively simple, but many are so difficult and expensive as to determine the very feasibility of the project itself.

The scheme of river control is an inseparable part of the construction program. It will usually determine the order of procedure; often features of design, and methods of construction. Too much emphasis can not be laid on the necessity for evolving a logical, systematic, practicable plan of stream diversion. If the foundation is adequate, a correct diversion plan adopted, and the masonry brought above the river bed, the remaining construction of the dam is usually simple.

The student is assumed to have obtained a knowledge of cofferdams from the course on foundations.

Access to the site, and topographical conditions determine that construction will be carried on from the east bank. The location of the structure is shown on Dr. E2-24. It is necessary to excavate an approach channel to the lower lock in the sandstone bed of the river, 7 ft. below lower pool level. This cut follows the flare of the lower guide walls and runs out 1000 ft. down stream from the lower end of the locks (measured in line with the lower river guide wall).

1. Before deciding on any order of construction or diversion scheme, what engineering and financial decision must you make?

a. What factors enter into your determination?

b. In the specific problem under consideration what decisions will you make?

2. Determination of the order of building the masonry structure is inseparable from the plan of river control. Submit a brief report describing the sequence in which the dam and locks will be built, and your general plan of unwatering the site. On Dr. E2-24 sketch in colors (using at least 3) the progress of the masonry. Show the location of the diversion structures to be built in their successive order. The program should be divided into not more than 4 steps.

a. How and when will the diversion works be removed? Consider this carefully.

3. Draw a sketch design of a 100 ft. section of the cofferdam.

4. Rail communication has been established. Standard gauge dinkeys, dump cars, track, a steam shovel and locomotive crane are available. How will you build the cofferdam? Illustrate by a free-hand sketch.

5. Costs of Material at Site:

(Insert correct unit prices)

Southern Pine No. 1 common	M. B. M.
60 lb. relaying rails and fish plates	ton
R. R. spikes	lb.
Round iron fabricated	lb.
Bolts and nuts	lb.
Round timbers average 16-inch diam.	lin. ft.
R. R. ties	each.

Installation Costs:

Labor	10-hr. day
Lumber, including nails	M. B. M.
Round timber	lin. ft.
Laying track on cofferdam	lin. ft.
Broken stone, in place	cu. yd.
Clay in place	cu. yd.

Make a bill of material and net estimate of cost of the diversion works you employ. Do not go into minute detail.

6. When cofferdams are destroyed by high water, the damage usually occurs at the time of over-topping, due to the washing effect of the falling water.

   a. What provision can be made to minimize this danger?

   b. On a sketch indicate method for minimizing the danger at the time of over-topping. (No details.)

7. Assume that the river bed at Squaw Shoals is 400 ft. wide between low water lines, the other conditions remaining the same. On drawing E 2-24 sketch the scheme of river diversion which will be used. Show all cofferdams and waterways.

   a. In this width of channel would it be practical to protect against all floods?

   b. Sketch a diagrammatic cross-section of the cofferdam neglecting all calculations as to stability but showing the special features which must be incorporated. No details.

8. Assume all conditions are the same as indicated in Section V, except that the banks are hard rock and precipitous. What plan of river diversion would be used?

On stream of small average flow, the permanent outlet works are often designed to be placed in the first section of the dam, and used to pass the ordinary flow after the foundation is in. The construction program is often determined by this consideration.

MINOR PROBLEM NO. III—EXCAVATING THE FOUNDATION

From borings, if properly taken, a good knowledge of the foundation should be obtained, and some idea as to the amount of unsatisfactory material it will be necessary to remove. The final decision must be reserved until the foundation pit is opened, and inspection possible. The foundation history of the dam under consideration is an excellent illustration of this.

The study of the excavated foundation must extend over a sufficient area to furnish certain proof that the condition observed is not local. This does not mean that the whole foundation must be opened before concrete is started; in fact, due to the plan of river diversion this is often impossible. The excavation should be studied in connection with the borings, and if the two are consistent, masonry can be safely started after a substantial percentage of the excavation is opened.

It is not the intention in this course to go into the methods of excavation, as the student is assumed to have some knowledge on the subject.

1. On cross section paper construct sections of the foundation as follows: Refer to drawing D 2—5 and R. R. 5.

   a. Section on line 20—00 assuming borings H—12.6 and W—23.5 are on this line.
   Between horizontal limits H—12.6 to W—23.5.
   Between vertical limits—195 to 300.
   Scale vertical 1 in. 210 ft.—0 in.
   Scale horizontal 1 in. 250 ft.—0 in.

   b. Section on line P—00, assuming borings P—05, Q—48, N—50, and Q—10 are the same line.
   Between vertical limits, 160 to 180.
   Between horizontal limits, 18—20 to 23—00.
   Scale vertical 1 in. 10 ft.—0 in.
   Scale horizontal—1 in. 50 ft.—0 in.

Study the consistency of the data, remembering that the borings were taken in flowing water, where there was a possibility of not recovering any soft soluble core. Soft cores are often not recovered at all.

c. Study the borings on line K—20. Are they consistent, and do they represent the probable condition exactly?

d. Make the same study on the sections you have constructed.

Do these sections represent the probable condition?

2. When constructing dams upon rock foundations, it has always been considered good practice to sink the masonry into the rock some distance, even if the character of the rock be satisfactory.

On the same cross section sheet construct a section on line P—00 as follows:

Horizontal limits, 20—00 to 21—50.
Vertical limits, —166 to —200.
Scale—1 inch=10 ft.

Sketch the accepted profile of the dam, placing the crest at station 20—35. Show how deep you will carry the foundation and the cut-off wall. Consider three things, i. e., foundation, pressures, disturbances of rock due to excavation and underflow. The question of the acceptability of this foundation requires careful study and thought.

a. Note that there is a solid sandstone layer 8 feet to 12 feet thick, across the bed of the river. Is this layer suitable for the foundation considered from the standpoint of strength?

b. In excavating rock in strata, great care is usually necessary in order not to shatter the whole layer. Fractures from blasting usually extend through and often loosen the entire stratum.

Is it advisable under the conditions existing in this foundation to attempt to sink the dam any considerable distance into the top layer of sandstone?

If not, how would you treat the surface of the rock?

c. If it is inadvisable to sink the whole dam into the rock, how may the design be made so as to increase the factor of safety against sliding?

d. Would the foundation be water-tight when subjected to a head of 63 feet? Give reasons.

e. If from examination of the borings you conclude that the foundation would possibly not be water-tight, how can you make it water-tight without sacrificing or disturbing an appreciable part of this solid sandstone layer?

f. Note that the sandstone of the river bed rises to a berm on each side of the river. The surface of the pool will be at El—224. Will your decision as to the adequacy of the foundation and depth of excavation be the same all over the site?

3. Making an excavation for the foundation of a dam requires the exercise of considerable judgment in the use of explosives, which in turn depend largely on the character and stratification of the rock encountered. The object to be obtained is to have a solid undisturbed rock bed to build upon and against. This is particularly desirable on the up-stream side and bottom of the cut, as it is here that percolation should be intercepted. In fact, modern practice indicates that the foundation down stream from the cut-off should not be water-tight.

a. An air plant and drills are available. How would you proceed to excavate the back edge of the foundation to avoid shattering the rock? Draw a free-hand sketch.

b. A cut-off trench 24 feet deep, bottom width 6 feet, is to be excavated through stratified rock. Jackhammer drills and 9 foot drill steel are available. Sketch the method of drilling to avoid shattering on the faces of the trench.

c. Channeling, a method much used in quarry work, is sometimes applied to construction where the rock is stratified and it is particularly desired to obtain faces and beds free from shattering. A channelling machine, by means of a rapidly striking chisel, cuts or chisels a groove in the rock about 3 feet to 4 feet wide, and to any desired depth and angle. It is not economical to operate beyond a depth of about 8 feet. For depths exceeding this, it is customary to break the cut into two or more lifts. View No. 367, (furnished the student), is an excellent illustration of the two methods of obtaining a clean cut in rock. The high up-stream side of the cut-off has been "line holed" and shot. The down-stream face against the foundation bed, where it is essential that no shattering take place, is being channeled in two lifts. The rock in the cut-off can be shot out with no disturbance to the foundation.

The steam channeler shown, while much used, is not now being manufactured. It has been superseded by the much simpler and more economical electric machine. The steam machine had the added disadvantage in marble quarries of staining the marble product.

The performance of the Standard 7 in. electric channeler varies, of course, according to the kind of rock to be cut. Observation was made in a marble quarry in Vermont, when the average length of cut was about 25 feet and 5 feet to 13 feet deep. A 7-in. machine cut an average of 10 square feet per hour. Much of this cutting was on an incline, otherwise the cutting would have been from 25 per cent to 50 per cent more. In the ordinary rocks encountered in construction work the performance will be higher than this for the actual running time. At least 25 per cent lost time must be allowed for changing tools and other unavoidable delays. The three channelers employed on the Black Warrior Dam average 100 square feet per day.

MINOR PROBLEM NO. IV—CUT-OFF AND DRAINAGE

Of the many indefinite features of dam design there are none on which there is less definite information and agreement among engineers than the subject of uplift. It is clear that the maximum of pressure which can exist on any horizontal plane in the dam or foundation is that due to the whole head of water above the plane at the heel decreasing to the pressure of tail water at the toe, and the minimum is zero where all water is excluded from the plane. The truth is probably well below the average between these two cases. The assumptions made by engineers likewise vary between wide limits. For the design of the Black Warrior Dam No. 17, the full head is assumed to act over two-thirds of the base area, while in the design of the Elephant Butte, the latest and one of the greatest of the world's high dams, water pressure was assumed to be two-thirds of the reservoir's head at the heel, diminishing uniformly to zero at the toe.

As a general proposition, there is much less danger of uplift within the masonry structure than there is within the sub-foundation. Concrete masonry should be much more impervious than any ordinary country rock as it lies in its natural bed. Obviously the greatest danger of uplift is at the junction between the masonry and the foundation rock.

There are two general theories and methods of preventing or reducing danger from uplift, one as old as the construction of masonry dams, the other a new, radical and important departure in dam building, the extension of which will probably produce great economics in construction.

The rock foundation at the Black Warrior Dam No. 17 was found to be remarkably homogenous and tight. No special construction methods were necessary. Further, the dam is not high. The difficulty of preventing uplift increases much out of proportion to the head of water.

The drawing furnished is a cross section of the Elephant Butte Dam, which was built on the Rio Grande in New Mexico.

1. What are the two general methods of preventing uplift in masonry dams? Discuss them.
2. On Drawing furnished, sketch the first method. Assume that the foundation is on broken and fissured rock.

   a. Discuss the results to be accomplished in building the concrete masonry.
3. On Drawing furnished sketch the scheme for the second method.

   a. Explain fully the difference in placing the concrete in this method.

   b. Show how you will treat the foundation, both at the surface of contact between the rock and masonry, and below the rock surface.

   c. In the new principle of dam construction what two results must be obtained in the sub-foundation?

   d. Explain just how these results will be obtained and the order in which the work must be done.

   All the necessary data as to assumptions and dimensions are shown on the profile drawing of Elephant Butte Dam. Recalculate the form of the dam, neglecting all up-thrust and masonry flotation. Superimpose this new profile on the old, using the same up-stream batter.

   f. Under the conditions shown on the drawing would it ever be physically possible to have zero uplift? Discuss this under both old and new theories of dam construction.

   g. Calculate the uplift per linear foot at the base under the following cases, assuming the physical conditions as shown on the drawing, with the reservoir full.

   (1) Dam foundation tight over the whole area.

   (2) Dam to have an impervious cut-off at the heel and an effective system of drainage.

   (3) When grouting rock it is better practice not to drill the grout holes in close proximity and succession across the cut-off or foundation. Much better results are obtained

by drilling and grouting to somewhat separated intervals, say from three to nine feet apart, and treating intermediate holes later. This may save much expensive drilling, as the intermediate holes may prove the foundation to be tight. The spacing of holes will depend entirely upon the character of the foundation and the results which which can be obtained by grouting. On the Drawing No. — of the Elephant Butte Dam, indicate by numbers the successive five stages in the grout treatment of the up-stream impervious zone, beginning from the open foundation before any concrete is placed.

MINOR PROBLEM NO. V—FOUNDATION MASONRY

The foundation of the Black Warrior Dam No. 17 proved to be remarkably tight and dry. In order to develop some of the principles of placing masonry another foundation will be assumed.

The Tremp Dam, built on the Noguerra Pallersa River, Spain, presented some interesting foundation developments. The general layout is shown on drawing furnished.

The river is a rapid flowing mountain stream, fed by the snows of the Pyrenees. As is the characteristic of most rivers of this nature, it is "flashy," being subject to great and sudden rises any time, and has well defined seasons of high and low water. No reasonable diversion works could be built which would pass the floods during the high water season without flooding the pit.

At the dam site, bedrock was covered with from 10 to 20 feet of loose overburden, which gave great trouble by washing in when the pit was flooded. Bedrock was of excellent quality, but found to be crossed diagonally to the axis of the dam by several large, wide faults, almost vertical, and of indefinite depth. Examination of the bedrock indicates the necessity of a deep cut-off. The foundation was rough, pitched sharply to the lowest point near the upsteam middle of the site, and was quite wet. The concrete plant consisted of a battery of four one-yard mixers, delivering the concrete through chutes and capable of placing more than 100 yards of concrete per hour. A plan and cross section of the foundation in the river bed is shown on drawing furnished. Ample compressed air was available.

1. By the time bedrock had been stripped it was found that the "low season" was so far advanced that if time were taken to excavate the cut-off and the fault the masonry could not be brought above high water. This would mean washing a great amount of loose material into the pit, and probably the loss of several months' time.

   A difficult proposition is presented. Draw a sketch illustrating two methods of procedure which can be followed to save the situation. One method is much the better, and was followed. Be prepared to discuss your solution.

2. When dealing with inflowing water the head of concrete and the volume which can be placed quickly is of great importance. A large volume and particularly a high head of fresh concrete, quickly placed, will be effective where an equal volume, slowly placed or poured in low, wide forms, will be honeycombed and fail to accomplish the result.

It is essential that the dam be as tight along the back or heel zone as construction methods can make it. Modern practice is to drain the foundation down-stream from this impervious zone. "Blind" or "Roman" drains, consisting simply of a line of loosely piled hand stone or crushed stone, are often effective.

The object is usually so to plan the work that while the concrete is fresh the water shall have a free escape and can be afterward effectively sealed. This is not always possible, particularly when closing the foundation bed of masonry over the lower part of the site.

a. Study the drawings and determine where you would start the masonry and how you would proceed.

b. Discuss several ways to be used to relieve the water pressure.

c. Assume that the water pressure is coming in through sub-foundation cracks at a head of 10 feet. When smothering water it is important to remember that speed is of prime importance. What dimensions would you make the forms?

d. Would it be necessary to make the forms the same size over the whole wet area, and when the concrete is being closed?

e. Assume that the water has been crowded into one part of the foundation and the surrounding masonry brought up to the height greater than the static head of the water. The inflow is so great as to make the deposition of concrete without washing impracticable. How would you build this portion of the foundation? Draw a sketch of your apparatus and describe its use, and the precautions to be observed.

f. What precautions would you take at the mixer plant when closing?

B. SYNOPSIS OF THE COURSE IN MATERIALS
TIMBER

First comes a general consideration of the classes of trees and applications (3), with structure (7), inspection and defects (7), decay (4), seasoning (8), mechanical properties (8), joints (2), classification of timber (8), mill construction, costs, and specifications (7), preservation (9), identification (3). Numerals in parentheses indicate the number of pages of typewritten matter. The text is a well-balanced summary of facts relating to wood technology.

The following books are placed in the students' hands, to be used as reference books only—not as text-books:

Materials of Construction—Johnson.
American Civil Engineers' Handbooks—Merriman.
Civil Engineers' Pocketbook—Trautwine.
Architects' and Builders' Handbook—Kidder.

PROBLEM No. I.

Title: Design of wood framework for a warehouse at Belvoir, Va.

Situation: Material is partly available from reclaimed material now at post. Size of building, specified with overhead clearance, live load, wall thickness.

Job: Student is to prepare a general design with specifications and an estimate of the cost. He is to fix the layout of columns and to decide upon fiber stresses. He is to select the species of wood for the various elements, design the joints, stairways, provide for fire protection, decay resistance, finish, furnish a bill of material, choose the type of roof, and design it.

Subsidiary Situations:

(1) If warehouse were to be in Denver, what new choice of species, and how would this affect the sizes?

(2) If 12x12-inch timbers for columns were condemned by inspector could 6x12 pieces bolted together be used? If so, how?

PROBLEM No. II.

Title: To choose ties in Arizona.

Situation: Seven species available; costs given, plain and treated.

Job: To choose ties and give reasons; specify methods of preservative treatment.

Brings in Wood Preservation and Track Construction.

PROBLEM No. III.

Title: Pipe line for Water Under Pressure at Atlantic City.

Situation: A pipe line under head of 10 feet is to be built over tidal flats to Atlantic City, passing under one tidal inlet.

Job: To choose material for pipe, discussing factors governing decision and give reasons. Location of line, above or below grade.

Brings in pipes and siphons. Strength and durability of wood.

PROBLEM No. IV.

Title: Material for Pontoons.

Situation: Proposed changes for material for heavy equipage pontoons are to be investigated.

Job: To report on best materials for various parts of pontoon. Brings in strength and durability of wood and pontoon construction.

PROBLEM No. V.

Title: Timber Pier at Belvoir.

Situation: A pipe line under head of 110 feet is to be built or 300 lbs. per square foot.

Job: To design pier.

1. ANALYSIS OF PROBLEM No. I INTO MINOR PROBLEMS

A permanent warehouse is to be built at Belvoir, using wood framing throughout. It is assumed that materials are partly available from reclaimed material now on the post, but that where necessary new material will be used.

The building will be 59′-0″ x 245′-0″ from outside to outside of brick walls. Because of the intention to use overhead trolley conveyers, the minimum clearance in stories will be:

Basement, 9′—0″
First Story, 16′—0″
Second Story, 14′—0″
Third Story, 14′—0″
Fourth Story, 12′—0″

Live load on all floors, 175 lbs.

Wall thickness:

Basement and first story, 21″
Second and third story, 17″
Fourth story, 13″

Two different types of roof are under consideration. The first continues the column construction clear through to the roof; in the alternate construction the fourth floor is to be clear of columns.

The student is assumed to be the Officer in Charge of Design and Construction. He is to prepare a general design, with specifications and an estimate of cost of the wood framing (the matter of cost may be confined to an estimate of the cost of one typical floor).

I. (a) What is the best construction in timber, and what are its principal features?
(b) What are the primary considerations governing the layout?
(c) To what extent should the first cost govern the layout?
(d) Shall we lay out the width or the length first?
(e) Which way shall the beams run?

II. (a) What are the governing conditions for allowable fibre stress?

(b) What fibre stress can be used?

III. What wood shall we use?

(a) For beams.

(b) For girders.

(c) For columns.

(d) For sub-floor.

(e) For flooring.

(f) For sheathing.

(g) For sash.

(h) For frames.

IV. (a) How shall we space the columns transversely?

(b) What would be the ideal spacing?

(c) Compare wide side spans and a narrow center span with a number of equal spans.

V. (a) What are the governing factors in longitudinal spacing of columns?

(b) How shall we space them?

VI. How shall we space the beams framing into girders? What is the most economical spacing?

VII. Sketch a section of the roof at the cornice line (the purpose of this question is not to show any brick details, but to show the necessary structural features at this point).

VIII. (a) What will be used at wall-line for bearing?

(b) What factors enter into the design at this point?

(c) How will we detail to secure all the desired points?

IX. (a) What shall we use where a beam frames into a girder?

(b) What factors enter into the design at this point?

(c) How shall we design to secure the desired points?

(d) Are there any other devices that might be used?

X. (a) What shall we use when the girders and beams frame into columns?
(b) What are the limitations of our materials?
(c) Can we overcome them satisfactorily in our design?
(d) Are there any other methods of construction that could be used here; if so, what are their advantages and disadvantages?
(e) Can you rest the columns on girders?

XI. (a) What will you use to carry the floor at the wall-line?
(b) What dangers of deteriorating factors exist here?
(c) How can you avoid them?
(d) Is it customary to take these precautionary measures?

XII. What will prevent the walls from falling outward?

XIII. (a) Assuming concrete footings of a value of 350 lbs. per square inch, what can be done to supply sufficient bearing for the basement columns?
(b) What fibre stress can be used here?
(c) What bearing material would you use?
(d) How will you anchor the bases down?

XIV. (a) What will you use for general fire protection?
(b) What materials will you use for containers? Why?
(c) How will you support it and on what materials?
(d) Give reasons for your choice?

XV. Write a bill of material for the 3rd floor. Estimate its cost.

XVI. What paint or other covering will you use on the timbers?

XVII. (a) How will you eliminate dry rot once it gets started?
(b) Is it necessary to keep the tops of the beams and girders at the same level?
(c) What would you do if the girders were of steel and the beams were of timber? How would you support the beams if the girders were of steel?
(d) How would you support joists if the girders were of steel?

(e) What materials are available for use as columns?
(f) Is a wood column liable to progressive failure?
(g) Discuss reasons for your choice of column material.

XVIII. There is some 10-in. x 14-in. material on the post that possibly can be utilized for beams and by re-sawing can be used for columns. Write instructions to non-coms. of ordinary intelligence, but without any experience in timber inspection, by which they can select suitable pieces and reject unsuitable material.

XIX. Will you use dressed or rough lumber?

XX. (a) Shall we treat the timber in any way?
(b) How do you prevent rot from starting?

XXI. (a) The office is to be paneled in an ornamental wood. What will you select?
(b) How will you specify it?
(c) Will you use veneered or solid wood?
(d) If your choice is for veneered lumber, how thick will the veneering be?
(e) In what direction will the veneering be cut?
(f) How will you prevent shrinkage which might ruin the appearance of the office?

XXII. (a) How much slope will the roof require?
(b) What are the determining factors?

XXIII. (a) Is it necessary to use a fireproof stairway?
(b) Of what material will the stairway be composed?
(c) What will you use for stringers?
(d) For treads?
(e) For risers?
(f) For platforms?
(g) For railing?

XXIV. (a) Through an error on the part of the inspector at the mill, you find that it is necessary to condemn eight 12"x12" pieces which had been shipped for use as first-story columns. Name five probable defects, any one of which may cause rejection. The foreman suggests that inasmuch as it will take three weeks to get new columns from the Southern mill that some of the 6"x12" purchased for girders shall be bolted very well together and used in columns, thus enabling the work to proceed. What is your decision?

(c) Write a bill of material showing the urgent requisition which you will put in to avoid tying up the entire job.

XXV. (a) How much clearance will be required between girders, beams and columns?

(b) How will you allow for expansion and contraction in height?

(c) Is this matter ever neglected in buildings of frame construction?

XXVI. Suppose the building were in Denver, instead of Belvoir? What would be your choice of material?
What effect would it have on sizes of material?

XXVII. (a) What type truss will you use for roof?

(b) Design it.

(c) Compare its cost, in place, with the cost of using a type of construction in which the columns are carried up to the roof line.

(d) Dctail: The heel joint,
peak joint,
a tension splice, differnt from any shown in the pamphlet on "Timber."
An intermediate wet member joint.

2. INSTRUCTOR'S NOTES ON MINOR PROBLEM No. I.

INSTRUCTOR'S NOTES—The notes which follow show the answers to Minor Problem No. 1, and are typical of the answers which have been prepared for all questions on all problems. By means of these notes to instructors, it is expected that uniformity of instruction will be secured, in spite of changing personnel; and a permanent record is secured of the opinions of experienced engineers, which record will always be available for the use of future instructors.)

Question I-a.

Direct student to use "slow burning" or "mill" construction. Discuss the general features which constitute this type:—it consists in a construction of timbers in large, solid masses so as to expose the least number of corners or other easily ignited points; also with a view toward facilitating every point being easily reached by water in case of fire.

Mill construction consists in isolating every floor by means of non-combustible fire stops.

All especially hazardous points, whether such points are caused by stock or operations, are guarded by fireproof materials, such as asbestos or plaster or metal lath.

All suitable safeguards against fire are provided.

Floors must be kept as free as possible of combustible partitions, with the idea of preventing the building from being split into a number of separate cells, in any one of which a fire may attain a good start before detection.

Timbers and floors must be of sufficient size.

Walls should not be furred. Highly inflammable varnishes must not be used.

Windows exposed to fire hazards from adjacent buildings must be protected by wire glass and fire shutters.

A minimum of wood must be used for interior finish.

No beams less than 6″ wide may be used. Flooring must be at least 4″ thick, laid either with or without a hardwood floor above, roofs to be at least 3″ planks. The floors and roofs will be grooved and splined. The grooves will be 7/8″ x 7/8″ and the splines of pine 7/8″x1¾″. The edge of the floors will be kept ½″ away from the face of the brick walls to avoid any danger of cracking the walls by the floors swelling. The open joint will be covered by a batten above and below. Between the sub-floorings and the over-flooring will be laid two layers of building paper, breaking joints and mopped with hot tar. The over-floor will be laid at right angles to the sub-floor. The sub-floor planks will be considered as acting simply as beams without any allowance for either continuous beam or slab action.

Question I-b.

The general layout of the building is governed by:

(a) First cost.

Since we are dealing with timber construction, the fact must be emphasized that with this material the owner pays for his timber in lengths of even feet; whereas, in steel and concrete, he pays only for the actual length that goes into the building. For instance, if a beam 18′ 3″ long is required in timber, a piece 20′ long must be purchased. The first cost of any timber construction, therefore, is dependent greatly upon the skill of the engineer in laying out so as to avoid excessive waste.

(b) The layout is dependent upon the direction from which light comes; a minimum of obstructions to the passage of light must be used.

(c) The layout must be such as will facilitate best the use of the building for its intended purpose. Thus in a machine shop it may be necessary to provide clear floor space, even though this result is attained at a considerably increased first cost; on the other hand, the warehouse may have columns scattered over the floor, provided they are not so numerous as to interfere with the easy trucking of loads.

(d) It will be brought out that in any "slow burning" construction the combined area of all columns is an amount that is directly dependent upon the weight of an entire floor; but note that the larger the number of columns shown on the transverse section of the building the smaller the portion of load carried by the walls; and hence, the more column area that will be required. It, therefore, follows that in using wood the judgment of the engineer must be exercised to reach a point where the column area is at a reasonably low figure without causing the size of the beams and girders to be unduly increased owing to the long spans which would result from an insufficient number of columns.

Question I-c.

The adaptibility of the building for the use for which it is intended must always be the final determining factor in the layout, and every other consideration must be subordinated to that. Thus, a certain layout might involve an increased cost of $1,000.00; but it is not at all difficult to conceive the cheaper layout as involving an extra operating expense in the unnecessary handling of materials or inconvenience to workmen that would amount to $1.00 per day, and this would mean $300.00 per year, or 30 per cent, which might be saved by spending the additional thousand dollars in first cost.

Question I-d.

We shall lay out the widths first. This is the usual custom, as long experience has proven that after an economical layout in width has been secured the longitudinal spacing of the columns can be easily adjusted. The greatest amount of travel in a building is in the direction of its length, and it is important that the passageway between columns be adjusted with a view to a minimum of obstructions to lengthwise travel.

Question I-e.

We shall assume that the beams will run transversely and the girders longitudinally; it will be necessary that the girders have a shorter span than the beams if they are not to be of excessive size. To run the beams longitudinally and the girders transversely at once commits us to the use of three lines of columns spaced about 14′ apart, which apparently is an undue waste of floor space.

If we find that our assumptions for laying out the width first and for running the beams transversely have carried us into an extravagant construction, we will have to make an alternate design. If sufficient time were available it might be well to have the students lay out the floor, using the contrary assumption; time will probably not be available. Various layouts have already been thoroughly gone over by the instructors for the course and it has been found that the assumptions in question I-c and I-d are correct.

CHART INDICATING RELATIONSHIP BETWEEN TYPE OF INSTRUCTION AND PRIMARY PURPOSES

IN

Course in Mechanics of Engineering, at the Engineer School, Camp A. A. Humphreys, Va.

EXPLANATORY NOTE—These charts were prepared, after the course had been given, with a view to setting forth the principal objective in this type of instruction and its relation to the material of the course. The material in the body of the chart necessarily does not differ widely from that found in any well taught course, under any system; but it must be borne in mind that the horizontal sequence in the chart is extremely significant, and that the grouping of column headings is deliberate and not accidental. Before a principle has been stated to the student he is plunged into the analysis of the problem and only becomes cognizant of the principle as he needs it and applies it in his work. This sequence is chosen with the definite intention of accomplishing the primary aim of developing the student by means of the activities into which he is forced by the detailed analysis and discussion of a real engineering problem. The principal phases of classroom effort naturally fall under columns 6, 7 and 8. No attempt is made to teach engineering specialties, but the fields of these specialties are used as a means of broadening the student's view while he is being given a general development and is being taught mechanics of engineering. During the earlier phases of the discussion of each problem, the special features pertaining to that type of problem and which must be considered in a complete investigation, are discussed; but the instructor necessarily directs the main channel of effort into those considerations which pertain more directly to the subject of mechanics. The mechanics phases only are outlined in the chart. It is a skeleton outline and shows neither the relative time spent on the various parts nor the entire field covered by the discussions.

STUDENT ☞			INTEREST aroused and maintained by contact with Engineering Situations INDEPENDENT thought and judgment stimulated by analysis of new Problems			EFFECTIVE HABITS of work and study, and PRACTICAL use of working tools established		ACQUAINTANCE with Fundamental Principles, and KNOWLEDGE of their relations to solution of Problems		CO-ORDINATION and Fixation of Ideas	OUTLOOK Expectancy concerning future studies	
SUBJECT MATTER ☞			MOTIVATION Project or Problem	The SETTING—A concrete engineering situation		THE ANALYSIS		PRINCIPLES		Rules, Formulas, References, Data, Text	Deferred Analysis	Adv. Theory and Research
				As a Whole	In Part	Process	Working Tools	Basic	Derived			
1	2	3	4	5	6	7	8	9	10	11	12	13
S-1	A		Military necessity requires transportation of 12" howitzer	Across King Post Truss Highway Bridge of 41' span, over gorge 100' deep with very steep rocky sides. Bridge 5 years old and in good condition	Is it safe?	Make list of possible causes of failure, (be exploratory only)	Working drawings. Previous experience and judgment.			*		Material for advanced theory and research appears in subsequent problems
					Probable effect of load on the bridge	a. Trace loads from point of application to the ground.	Previous experience and judgment.					
						b. Effect on various members.	Sketches of possible failure.					
	B	1	Floor sys. Safety of Planking	4" Plank on 4"x12" Stringers 2'-0" c.c.	Failure by bending	Position of load to break	Sketches.		Physical idea of bending		Bending Stresses	
		2	Stringers	4"x12"x12', Supported each end	Failure by bending	Position of load to break	Sketches		Physical idea of bending		Bending Stresses	
		3	Floorbeams	2-8"x14" Supported each end by hanger	Failure by bending	Position of load to break	Sketches		Physical idea of bending		Bending Stresses	
	C	1	Truss. Safety of Hanger	1" round steel bar (not upset)	Determination of effect of dead load	Calculation of amount and distribution of dead load	Working drawings. Handbook; calculating amounts and weights of materials		Proportional Distribution of Loads	Handbook data for timber and steel		
					Determination of effect of live load	Position of amount of live load giving maximum load on hanger. Impact.	Working drawings. Inspection and judgment.					
					Force tending to break hanger	External force resisted by internal stress	Free body sketch.	Conditions of equilibrium	2 forces equal and opposite, on same action line	Rules for drawing forces in F. B. Sketches.		
								Forces are vector quantities and may be represented by lines.	Force is known when a) Pt. of App., b) direction, c) sense, d) magnitude are known.	Definition of "[illegible]"		
					Is hanger safe?	a. Distribution of stress over cross-section	Tests of steel (tension)	Stress uniformly distributed	Unit stress = total stress divided by area of cross section			
						b. Determination of breaking strength of hanger	Test of steel (tension)	Material safe within proportional limit	Working Factor of Safety	Ultimate strength and prop. limit for steel		
					Design safe hanger in conformity with good practice	Determination of necessary area of cross-section	Specifications based on tests of steel and good judgment.	Stress uniformly distributed	Total stress divided by unit stress gives area of cross section			
		2	End Post	8"x8" wood	Is it safe?	a. Find forces acting upon upper chord—analytic and graphic solutions	F. B. Sketch upper joint. Force polygon upper joint.	Equilibrium	$\Sigma H = 0$ $\Sigma V = 0$ Graphic and analytic			
						b. Find apparent unit stress equals load/area. Why so large a stick?	Judgment. Experiment with various models of columns. Empirical Formula	Stress in long columns is combination of direct stress and bending	Strength of long columns is dependent on—length, shape of x-sec. and area of cross section, and material.	Specifications. Straight line formula.		
						c. Design end post in accordance with specifications	Standard formula					
		3	Lower Chord	1¼" Round steel bar (not upset)	Is it safe?	a. Find forces acting upon it.	F. B. Sketch of joint at end post. Force polygon of joint.	Equilibrium. Three forces.	$\Sigma H = 0$ $\Sigma V = 0$ Graphic and analysis			
						b. Find unit stress and compare with allowable	Given minimum cross sectional area and tests of steel	(Same as for hanger.)	(Same as for hanger.)			
						c. Design lower chord	Standard Specifications	(Same as for hanger.)	(Same as for hanger.)			

						b. Find unit stress and compare with allowable.	Given minimum cross-sectional area and [illegible]	(Same as for hanger.)	(Same as for hanger.)			
						c. Design Stress chart.	Standard Specifications.	(Same as for hanger.)	(Same as for hanger.)			
		4	Bearing of truss on sills	Sills 8"×6"	Is it safe?	a. Find pressure of truss on sill.	F. B. Sketch of entire truss. Simple [illegible] $\Sigma V=0$, $\Sigma h=0$		Proportional distribution of loads.	Handbook data.		
						b. Find unit pressure and compare with safe.	Test of bearing strength					
5-2	A	1	A water tank on frame tower and concrete foundations fell over in a high wind. Investigate and report why it failed.	Circular tank 18' diam. and 10' high. 50' tower consisting of timber frames—4 dim. posts are 10' apart at top and 20' at base. Wreckage made it difficult to decide what part failed first.	1. What causes contributed to the failure of the tank?	Make a list of the items which might have been responsible. [illegible]	Working drawing of the tank and tower including foundations. Judgment and experience					
					2. Tower may have tipped over on its foundations.	a. Find forces tending to hold tank in place, full and empty.	Drawing of tank. Determination of weights of materials.	A distributed force may be replaced by its resultant.	A system of parallel forces may be replaced by a single force acting thru their C. of G.			
						b. Find forces tending to upset the tower.	Tests of wind pressure on flat and curved surfaces.	Fluid pressure.	Pressure depends upon character and position of exposed surfaces and velocity of wind.			
						c. Which produces greatest effect? How do forces produce rotation?	Sum of forces and effect of forces acting on tower in general. [illegible]	Conditions for equilibrium.	Noncoincident moments about any pt. [illegible]			
						d. Compare condition of tank full with tank empty. What factor of safety? What precautions to prevent failure?	Judgment.	Conditions for equilibrium	Downward reactions require tension connections to footings.			
					3. Foundations may have failed thru excessive soil pressure.	a. Find total pressure exerted by each footing.	F. B. Sketch. Principle of moments. Graphic determination of reactions. String polygon.	Conditions for equilibrium	[illegible] Analytic and graphic			
						b. Find maximum unit pressure on soil. Is it safe?	Tests of safe loads on various soils. Judgment					
						c. If unsafe, how remedy?	Specifications—Judgment.					
					4. Tower may have been overstressed as a truss.	a. Find stress in columns.	F. B. Sketch—Laws of equilibrium. Force Polygon. Method of sections. Principle of moments.	The principle of structures [illegible] as a truss	If a structure is in equilibrium, any part taken as a free body is in equilibrium.			
						b. Find stress in bracing.						
						c. Find stress in struts.						
						d. How remedy weak members?	Specifications.					
					5. Braces in bent	a. Failure by bending.	Sketches				Bending Stresses	
						b. Failure by crushing on top of posts.	Allow. unit bearing stresses					
5-3	A	1	A design for a concrete dam has been submitted in connection with a proposed water supply system for a certain city located in the southern part of the United States.	The dam is straight, 500 feet long and creates a reservoir extending half a mile up stream from the dam. Trapezoidal Section {Top 5' Base 25' Height 30'. Elevation of crest of dam, 800' above sea level. Tests have shown the foundation to be solid granite.	1. Is dam safe? Investigate and report	List possible sources of failure. (As experience may)	Laws of equilibrium. Judgment and experience				Class of [illegible]	
					2. Will it slide on its base?	a. Find forces which produce sliding	F. B. Sketch. Hydrostatic prism		Hydrostatic Pressure			
						b. Find forces which hold against sliding	Experiments in friction.	Frictional force = some per cent of normal pressure.	[illegible]	Allowable coefficient of friction from handbook		
						c. Find ratio between a and b.	Judgment determines proper factor of safety					
					3. Will it overturn?	a. Find moments of all forces tending to overturn.	F. B. Sketch—Laws of equilibrium	Equilibrium.	Σ Ms about toe of Dam			
						b. Find mom. of all forces tending to stability	Ditto	Equilibrium.	Σ Ms about toe of Dam.			
						c. Find ratio of stabilizing to overturning moments	Judgment as to allowable Factor of Safety	Equilibrium	$\frac{\Sigma Ms}{\Sigma Ms} = F.S.$	Factor of safety should be [illegible]		
					4. Is the pressure too great on the foundation?	Find distribution of pressure on base. Find max. and min. pressures per unit area.	Experiments—judgment—straight line variation, an acceptable assumption. Equilibrium data.	The resultant of a system of two equal and opposite forces in same line does not change previous conditions of equilibrium.	Pressure is combination of Direct Stress and bending. Couples in same plane are equivalent if their moments are equal and in same direction.	Safe values taken from handbook.	Uplift due to water pressure.	
5-4-a	A	1	Dam for water supply	Concrete dam in 5-3	1. Is dam safe at any section between top and bottom? Investigate and report.	a. Analyze as in 5-3 for sliding at sections 6', 12', 18', 24', and 30' below top of dam	Make tables of sliding forces—sketch of hydrostatic prism for each section.		[illegible]	Drill in method		Structures other than horizontal section
						b. As in 5-3 for overturning about toe of each Section	Tabulate moments and factors of safety F. B. Sketches.	Equilibrium	[illegible]	Drill in F. B. Sketch and [illegible]		
						c. As in 5-3 find max. and min. pressure on each section [illegible]	1. Tabulate Direct Pres. Bending moments due to water. Bending moments due to eccentricity of dam, about center of base—Table of net bending mom. at center of base, F. B. Sketch of each section.	Equilibrium.	[illegible]	Drill in methods and F. B. Sketches. Drill.		

							[illegible] due to water. [illegible] moments due to eccentricity of dam, about center of base—Table of [illegible] bending [illegible] at center of base. F. B. Sketch of each section.			[illegible]		
							5. Tabulate combined effect of direct stress and bending	[illegible]	Combined stresses.	Drill in combined stresses.		
	B	1	Simplification of investigation	Concrete dam, S. J.	1. Coordinate calculations for sliding and bending moments—[illegible] to "A"	a. Construct diagram showing "[illegible]" of water pressure.	Hydrostatic press. Graphic representation of quantities.		Load diagram.			
						b. Above the load diagram construct diagram showing tendency of each section to slide off	Summation of forces on a section		[illegible] Transverse Shear diagram.			
						c. What relation between a and b?	Geometry and Algebra		1st and 2nd Laws of derived curves			
						d. Above the shear diagram construct diagram of bending moments or tendencies to "break off."	Use data obtained in "A"		Moment diagram.			
						e. What relation between d and b?	Geometry and Algebra		1st and 2nd Laws of derived curves.			
						f. What relation between d and a?	Geometry and Algebra		2nd law of derived curves.			
						g. Coordinate the relations between load—shear—moment.	Physical ideas of load, shear and moment		Shear—Rate of change of moment Load—Rate of change of shear			
		2	Derived curves on roads	Level notes of survey	Profile and rate curves	a. Plot profile			Elevation diagram.			
						b. Plot rate curve below profile	1st and 2nd Laws of Derived Curves.		Grade rate diagram.			
						c. From rate curve, find difference in elevation between any two stations.	2nd Law			Drill in Laws and use of derived curves.		
		3	Ditto	Vertical curve on R. R. [illegible]	Ditto	a. Plot profile and rate of change curves	1st and 2nd Laws					
						b. Use diagrams to calculate intermediate elevations	ditto					
						c. Profile approaches parabola, rate curve approaches st. line.	ditto					
5-6	A	1	Timber cofferdam	9'x30' inside sheet piling 4"x12" sheet piles 6"x8" wales 6"x8" struts 2 wales 6' and 12' below water [illegible] specifically.)	1. Is the cofferdam safe? (No calculation data)	Make a list of all possible sources of failure. Discuss collection of information.	Working drawings. Judgment and previous experience.					
					2. Investigate strains and report as to safety.	a. Determination of load	F. B. Sketch. Triangular pressure prism	Equilibrium. Hydrostatic pressure				
						b. Find allowable Stress in column	Standard specifications	Combined bending and direct stress	Empirical column formula	Drill in use of straight line column formula		
						c. Make recommendations as to strengthening	Judgment.					
	B	1	Timber under bending.	A. Sheet piling, 2 Supports. Trapezoidal loading.	1. Will the piling be safe under the effect of bending?	Find external bending moment of forces acting on a short [illegible] taken as a beam	Laws of equilibrium. F. B. Sketches. Laws of derived curves—Load shear and moment diagram.		See Equilibrium [illegible]	Drill in determination of max. shear and mom. in a beam and use of laws.	The previously deferred problem in bending has become a major problem.	
					2. Fiber stress due to bending	Cut beam transversely at any point and determine internal forces acting on fibers.	F. B. Sketch showing internal stresses. Laws of equilibrium	External Bending Moment = Internal Resist. Moment at any section	Max. fiber stress occurs at point of max. bending. Max. bending occurs when shear passes thru 0	Refer to specification for allowable fibre stress in bending		
					3. Longitudinal shear (tendency to split.)	a. Take short section of beam and determine forces acting on each face	Ditto	Ditto				
						b. Take section of a with lower face at neutral axis and find unit stress along this face.	Ditto	Laws of equilibrium	Longitudinal shear is distributed uniformly laterally over longitudinal section			
					4. Bearing on wales.	Determine area of bearing surface and load.	Specification					
				B. Wales as beam. 4 Supports.	1. A continuous beam.	Which is stronger, a series of simple beams or one continuous?	Models of beams of small cross section		Simple beams show greater bending		Analysis of Cont. Beam deferred	
					2. A series of single beams	Analyze for fiber stress and long. shear.	In same manner as for sheet piling above					

Date

STORAGE BATTERIES FOR TRUCKS

A JOB SHEET MANUAL

WAR DEPARTMENT
COMMITTEE ON EDUCATION AND SPECIAL TRAINING
WASHINGTON

4

PREFACE

One of the lessons taught by the intensive war training experience was that rapid progress in learning can be made when men are trained for definite jobs. Hence a better result will be secured from the technical instruction for the Reserve Officers' Training Corps if the product needed by the army is first defined in terms of the jobs and operations which an officer may have to perform in the conduct of his daily routine.

Such definitions will also assist college teachers in their regular work by supplying problems and other material which will enable them to connect their instruction with the broad field of application in which the mastery of scientific principles is objectively revealed. Instruction based on problems and job sheets of graded difficulty seem to offer as effective a means as can be found of developing the ability to do things in civilian life as well as in the army.

The job sheets herein printed have been prepared by Prof. L. W. W. Morrow of Yale University. In this work he has had the hearty cooperation of Mr. A. L. Pearson, Electrical Engineer of the Construction Division of the Army and of Mr. G. W. Vinal, Associate Physicist of the Bureau of Standards.

Further information on this subject may be had from a circular of information on tractor batteries, for use in storage battery maintenance stations, prepared by Mr. Pearson and published by the Bureau of Standards.

C. R. MANN,
Chairman, Advisory Board.

PREFACE

One of the lessons taught by the intensive war training experience was that rapid progress in learning can be made when men are trained for definite jobs. Hence a better result will be secured from the technical instruction for the Reserve Officers' Training Corps if the product needed by the army is first defined in terms of the jobs and operations which an officer may have to perform in the conduct of his daily routine.

Such definitions will also assist college teachers in their regular work by supplying problems and other material which will enable them to connect their instruction with the broad field of application in which the mastery of scientific principles is objectively revealed. Instruction based on problems and job sheets of graded difficulty seem to offer as effective a means as can be found of developing the ability to do things in civilian life as well as in the army.

The job sheets herein printed have been prepared by Prof. L. W. W. Morrow of Yale University. In this work he has had the hearty cooperation of Mr. A. L. Pearson, Electrical Engineer of the Construction Division of the Army and of Mr. G. W. Vinal, Associate Physicist of the Bureau of Standards.

Further information on this subject may be had from a circular of information on tractor batteries, for use in storage battery maintenance stations, prepared by Mr. Pearson and published by the Bureau of Standards.

C. R. MANN,
Chairman, Advisory Board.

STORAGE BATTERIES.

Electricity is in very general use in the Army, and the storage battery affords a convenient and portable means of obtaining an electrical supply for a mobile army in field operations. Generating stations are inadvisable for field service conditions, as they require prime movers, such as steam or gas engines, with consequent noise, smoke and weight—all detrimental features as compared to a storage battery. The storage battery furnishes a means of storing electricity in the form of chemical energy at one time and later using this energy as electricity where desired and when desired.

The Army officer in the field should know the principles and operating features of storage batteries and have a general knowledge of standard methods of maintaining, applying and repairing them. In connection with Army transportation and communication an officer may encounter emergency conditions at any moment involving the use of storage batteries as an aid in the solution of his problem.

The success of the Army in the field depends on the Service of Supply, and in this connection storage batteries are used as follows:

1. Munition depots use storage battery trucks and tractors for handling explosives and munitions. The munitions are transported to and from magazines and are loaded on trains or ships by means of the storage battery trucks. The enclosed electric motor and an absence of any flame make storage batteries essential for safety in such installations.

2. Motor Transport equipment includes storage batteries great numbers either for power or lighting. The varied supplies for an army are handled by truck and tractors which require many storage batteries for lighting, starting or power purposes.

3. Army posts and Field Headquarters use small gas engine driven electrical generators in combination with storage batteries for general light and power purposes.

Communication Systems for an army are of many types, but storage batteries are essential and useful with all types. In Radio Communication storage batteries are essential for all applications of vacuum tubes to receiving, sending and amplifying stations. In telegraph installations, storage batteries are used as a source of power. In telephone installations of all kinds, the storage battery is essential. In many other communication devices such as lights, ground telegraph, etc., the storage battery finds a useful application.

Storage batteries are thus seen to have a varied application in army communication, transportation and power supply. The course in electrical engineering, as given in American schools, generally treats the storage battery from the electro-chemical theory standpoint and often in meagre detail.

JOB SHEETS

Directions to Students.

The Job Sheets consist of brief specific directions for work to be performed. Questions are then asked on the details of the work. These questions are arranged in the order of procedure for doing the work and relate directly to the work as well as to the principles underlying the work.

The student should look over his Job Sheet very carefully before starting the work. He should obtain answers to the questions on the Job Sheet by actual work with the apparatus or by analysis of data taken on the apparatus. A correct answer to each question should be obtained, but books or the knowledge of the instructor should be used only as a last resort.

The Job Sheets are numbered and arranged according to difficulty of tasks and adaptability for learning storage battery practice. Each Job Sheet, however, is an entity and only those need be used which time and the type of instruction allow.

The Job Sheets are for truck and tractor batteries, but may be altered to adapt them for any type of lead or alkaline battery if the rating and data given to the student at the top of the sheet is changed to conform to the battery to be used.

JOB SHEETS

Job No.—Title.

1—Dismantle a lead battery for repair.

2—Assemble a lead battery after repair.

3—Mix electrolyte for filling lead batteries.

4—Mix electrolyte for filling alkaline batteries.

5—Operate and adjust a battery charging motor-generator set.

6—Operate and adjust a battery charging synchronous converter.

7—Operate and adjust a constant potential charging circuit.

8—Operate and adjust a constant current charging circuit.

9—Charge a lead battery by the constant current method.

10—Charge a lead battery by the constant potential method.

11—Charge lead batteries at abnormal rates.

12—Charge an alkaline battery by the constant current method.

13—Charge an alkaline battery by the constant potential method.

14—Charge an alkaline battery at abnormal rates.

15—Determine the condition and state of charge of a lead battery.

16—Determine the condition and state of charge of an alkaline battery.

17—Prepare a lead battery for dry storage.

18—Prepare a lead battery that has been in dry storage for service.

19—Prepare a new battery for service.

20—Determine the efficiency and characteristics of a lead battery.

21—Determine the efficiency and characteristics of an alkaline battery.

22—Perform the operation of lead burning.

23—Charge a battery from an emergency supply.

24—Answer some general questions about batteries.

25—Answer some service questions about batteries.

JOB NO. 1.

Dismantle a General Lead Batteries Co.'s Type BH-15, 12 Cell Truck Battery for Repair.

Manufacturers' Data:

220 ampere hours.
4½ hours discharge.
43 and 15 amperes charging rate.
1270–1280 specific gravity, charged.
1120–1130 specific gravity, discharged.

1. (a) Under what operating conditions should the user of a battery dismantle and repair it?

(b) Does the time taken to ship a battery to the factory for repair often make it imperative for the user to repair his own battery?

2. (a) Should a battery that is giving trouble be dismantled only as a last resort? Why?

(b) Can you determine the cause of poor operation of a battery by means of any tests?

(c) So long as a battery takes a charge, are the plates still in good shape?

(d) If the battery will not hold a charge, what is indicated?

(e) What information determined by tests indicates the battery needs dismantling?

3. (a) Assuming the battery needs dismantling for repair, do you need any manufacturers' data to do the job? If so, what data?

(b) Is all the necessary data on the battery name plate?

(c) Can you tell the number of positive and negative plates in the battery from the name plate?

(d) Why do you need any data about the battery before dismantling it?

4. (a) Can you predetermine by tests that a battery needs cleaning? If so, how?

(b) Can you predetermine by tests that a battery needs new positive plates? New negative plates? Both positive and negative plates?

(c) If a battery needs new plates, and you can predetermine

this condition, is it better and cheaper to rebuild the battery or to buy a new one?

5. (a) Should the battery be charged or discharged before being dismantled?

(b) Has the state of charge and the specific gravity of the electrolyte previous to dismantling any relation to the job? If so, what?

6. (a) Previous to dismantling a battery, what are the tools to be assembled and the new battery equipments to be prepared?

(b) Should the electrolyte be poured from the battery as the first step in dismantling?

(c) Is there any objection to leaving the electrolyte in the cells until the cells are taken from the case? If so, what?

(d) Do the weight of the battery and facilities for removing electrolyte without spilling affect the answers to (b) and (c)?

(e) Should the specific gravity of the electrolyte in each cell be recorded before dismantling the battery? If so, why?

(f) Should the old electrolyte be saved and used again or should new electrolyte be used in the repaired battery? Explain.

7. (a) Name two or three methods for removing burned cell connectors from the battery posts.

(b) Can the connectors be removed so as to have them capable of being used again?

(c) Can the connectors be pulled by means of a puller? If so, how is the operation performed?

(d) Would there be any advantage in boring out the connector over the post before using a puller? If so, what?

(e) Does the post have to be trimmed before reburning the connectors, depending on the method used in dismantling the connectors?

8. (a) Can a connector be removed very readily if no puller is available?

(b) In boring out a connector, what size and type of bit should be used? Why? How can the bit be centered?

(c) What is the objection to using a drill bit? A wood bit? A bit smaller than the top of the post? Would you use the same size wood bit as drill bit?

(d) How deep a hole should be bored in the connector and where should the center be located?

(e) Does the connector become slightly loose when the proper depth has been reached and can the joint between the post and the connector be seen when the bit is removed?

(f) After boring out has been completed, how can the connector be removed from the post?

(g) If a screw driver is used to pry up the connector, what precautions can you take to insure not breaking the cover on the cells? If sealing nuts are present, can they be used?

(h) Should the filling plugs be removed while boring out and removing connectors? If not, why?

9. (a) After the connectors are removed should the jars be taken from the case before removing the elements?

(b) Should the sealing nuts, if present, be removed before removing a cell from the case?

(c) Are the cells clamped in the case or fastened by compound?

(d) If a cell sticks can it be readily loosened by running a hot putty knife around the edge?

(e) In pulling a cell from the case, should both posts be held by tools? If so, what tools should be used?

10. (a) After a cell is removed from the case, how can the element be removed readily by one man?

(b) Why should the element rest on the top of the jar a few minutes when lifted from the jar?

(c) How can the cover be removed most readily?

(d) If a first insertion of a hot putty knife does not loosen the cover, what other agencies can be used?

(e) If a flame is used on the top side of the jar to help loosen the cover, is there any danger of an explosion? Why? What precautions should be taken against explosions? Against melting the top of the jar?

(f) If a cover sticks, is it advisable to remove all the cover sealing compound with the hot putty knife and to clean the inside edges of the jar?

(g) How does the operation of removing a double flange cover differ from that of removing a single flange cover?

11. (a) After the elements are removed from the jars, should they be dismantled or should the separators be removed first?

(b) Can the separators be removed best with the element on edge, on its side, or upside down?

(c) Should the separators be saved and used again?

(d) Can the separators be used again?

(e) Should stock separators be kept dry or wet? If wet, in water or in electrolyte?

(f) Should the jars be washed out with ordinary water or with distilled water?

(g) Can a special tool be used for removing separators? If so, what?

(h) When is it necessary to use a separator inserter?

12. (a) Can the positive and negative elements be separated readily?

(b) Can you distinguish the positive from the negative plates by appearance? By construction? By number?

(c) Will the appearance of the active material on each plate be affected by state of charge? State of sulphation? Length of life of cell? If so, how?

(d) If new positive plates are needed, what will be the appearance of the old plates?

(e) Does the sediment in the bottom of the jars come from the positive or negative plates?

(f) What is this sediment and why does it occur?

(g) What is the difference in construction of the positive and negative plates, and why is a different construction used?

(h) Are the new positive plates assembled as received from the factory?

JOB NO. 2.

Assemble a General Lead Batteries Co.'s Type BH-15, 12 Cell Truck Battery that has been Dismantled for Repair.

Manufacturers' Data:

220	ampere hours.
4½	hours, discharge.
43	and 15 charging rate.
1270–1280	specific gravity, charged.
1120–1130	specific gravity, discharged.

1. (a) Inspecting the dismantled cells preparatory to assembly, what would you do to repair corroded posts? Loose connection of a plate to the strap? Cracked covers? Cracked jars? Jars whose tops have been injured? Defective filling plugs? Short posts?

(b) Can all the repairs mentioned in (a) be made?

(c) What defects in (a) can best be handled by using new parts?

2. (a) What determines the allowable loss of active material before a new set of plates is required?

(b) Can you tell by inspection whether sulphated plates can be brought into condition by charging methods or whether new plates are advisable?

(c) Should sulphated plates be scrubbed or washed to remove the outside sulphation?

(d) What objection is there to using a brush or cloth to wash plate surfaces?

(e) If plates are buckled, can they be straightened? If so, how?

(f) If the active material is very hard, what is indicated?

(g) If the positive plates are porous and disintegrated, what does this indicate?

3. (a) Can you use the old separators again? Is this advisable?

(b) Can the perforated rubber sheets be used again?

(c) Should the wood separators be assembled while dry?

(d) Is it advisable to use a cracked separator?

(e) What side of the wood separator goes next the negative plate?

(f) What side of the wood separator goes next to the perforated rubber sheet? Why?

(g) How should the separators be assembled in the element?

4. (a) How should the battery case be treated before assembly?

(b) Should the case be washed out with some alkaline material? If so, why?

(c) Should the case be painted inside with asphaltum paint? If so, why? How many coats are advisable?

5. (a) How should the old jars be treated before assembly?

(b) How and why should the compound be removed from the inside and outside of the jar?

(c) Will wiping out the top of the jar with a cloth dampened by ammonia help in the assembly? If so, why?

6. (a) Should the plates be assembled and the cover and the sealing nuts be placed in position before the separators are inserted?

(b) Are any gaskets used under the cover on top of the posts? If so, why?

(c) Can any lubricant be used to advantage on the sealing nuts? If so, what kind? Why?

(d) Will battering a thread on the post lock the sealing nuts in position? Is this advisable?

7. (a) In sealing the jar how can the compound be heated?

(b) Should the jar be heated around the top to help getting cover in position in case the jar top is warped slightly?

(c) Will a hot putty knife help get a finished job of sealing? If so, how?

(d) Will the compound only come to the level of the jar top and will it all be removed from the outside of the jar? If so, why?

8. (a) How are the cells placed in the case?

(b) How are they fastened in place?

(c) Should they be sealed in the case with compound? Why?

(d) Should grease be put on the outside of the jars to help in the assembly?

9. (a) Should the posts be cleaned thoroughly before burning?

(b) Should the posts be trimmed in all cases?

(c) What can you do to a post that is too short?

(d) What is a good tool to use to clean posts? To trim posts? To clean connectors?

(e) With how much force may the connectors be tapped on the posts before burning? When should the connectors be put in position?

(f) Will acid on contacts affect burning? If so, how?

(g) How far below the top of the connector ought the top of the post to be?

(h) Is there any way to insure your getting a positive post always connected to a negative post by means of the connectors? If so, how?

10. (a) How is the connector burned to the post? (See Job No. 22).

11. (a) When is the electrolyte placed in the cells during the assembly process?

(b) What ought to be the specific gravity of the electrolyte used in the cells? Why?

(c) Does temperature have any effect on the assembly? If so, what?

12. (a) Should the battery be left standing after assembled? If so, why?

(b) How will the battery need to be treated to put it in operating condition?

(c) Why is the charge longer than normal?

(d) What modifications in the assembly would be made if the battery had bolted connectors instead of burned connectors?

JOB NO. 3.

Prepare Electrolyte for Filling an Electric Storage Battery Co.'s Type MVY-15, Ironclad Exide 12-cell Truck Battery.

Manufacturers' Data:

220.5 ampere hours.
$4\frac{1}{2}$ hours discharge.
49 amperes discharge.
40 and 12 amperes, charging rates.
1270–1280 specific gravity, charged.
1120-1130 specific gravity, discharged.
80° F. normal temperature.

1. Do you need any manufacturers' data in order to prepare the electrolyte? If so, where can the data be obtained?

2. Will data on electrolyte for any lead acid battery serve in mixing electrolyte for any other type?

3. (a) Can you determine the total quantity or weight of electrolyte needed for filling the 12 cells? If so, how?

(b) Can electrolyte be prepared in quantity and stored for use as needed?

(c) What materials can be used as electrolyte containers?

4. (a) What materials are used in making electrolyte?

(b) Where can the acid be obtained?

(c) What specifications are needed as regards purity and specific gravity of the acid?

(d) Will ordinary drug store C. P. acid suffice for making electrolyte?

(e) Why is it better to used distilled water than hydrant water in making electrolyte?

(f) How could you obtain distilled water in an emergency? Where is it obtained?

(g) In any case, can you use river water? Hydrant water? Rain water? Well water? Which is best to use?

(h) In case you desired a water analysis, where would you have it made?

(i) Is alkaline water worse than mineral water for use in mixing electrolyte? If so, why?

5. (a) What materials can be used in making the mixing vessels for electrolyte?

(b) In mixing, what proportions of acid and water will you use? By weight? By volume?

(c) Has the temperature, state of charge and specific gravity of the electrolyte previously used in the battery any effect on the proportions used? If so, why?

(d) Will you pour the acid into the water or the water into the acid? Why?

(e) Should the electrolyte be stirred while mixing and should it be allowed to cool before putting it in the cells? If so, why?

6. (a) Is all sulphuric acid of the same specific gravity?

(b) What is the meaning of the term specific gravity?

(c) What effect has temperature on specific gravity?

(d) What is a battery hydrometer? Why does it read specific gravity correctly?

(e) Why is the specific gravity of electrolyte prepared for filling new cells made about 1300?

(f) Why is the specific gravity made 50 points higher than that of the electrolyte previously used if new separators are put in an element?

(g) Why is it advisable to prepare new electrolyte with a specific gravity of about 1300 at normal temperature?

(h) Why should the specific gravity of the electrolyte in a cell never exceed 1300 or be less than 1100 at normal temperature?

(i) What would happen to the plates and to the ampere-hour capacity if either of the limits mentioned in (h) were exceeded? Why?

7. (a) Is it better to mix a reserve supply of electrolyte with high specific gravity or low specific gravity?

(b) What specific gravity would be used in filling a cell that had been in dry storage?

(c) That had been dismantled and cleaned?

(d) That had been dismantled and new positive plates inserted?

8. (a) What will be the effect on capacity and operation if a

charged cell is filled with electrolyte having a specific gravity of 1130 at normal temperature? Why?

(b) What will happen to the capacity of a discharged cell if filled with electrolyte having a specific gravity of 1280? Why?

(c) If newly mixed electrolyte reads 1265 at 100°F, what will be the reading at 80°F? At 32°F?

(d) Why does the specific gravity of the electrolyte change with the temperature?

9. What will you do with the newly filled battery to prepare it for service?

10. (a) To what height should the cell be filled with electrolyte?

(b) What occurs if the electrolyte does not cover the plates and the battery is placed in service?

JOB NO. 4.

Prepare Electrolyte for Refilling an Edison A6 Truck Battery.

Manufacturers' Data:

225 ampere hours.
5 hour discharge.
7 hour charge.
45 amperes normal discharge.
45 emperes normal charge.
4.3 pounds electrolyte per cell. (5% allowed for loss.)
21 cells.

1. (a) What materials are used to make up the electrolyte for Edison batteries?

(b) Can they be purchased at drug stores and prepared in a charging station?

(c) Should the electrolyte be obtained from the factory in nearly all cases? Why?

(d) Is the electrolyte shipped already prepared?

(e) Is it ever shipped in a dry, powdered form? If so, under what conditions?

2. (a) Why does the electrolyte deteriorate with use? What is the evidence of electrolyte deterioration?

(b) Has the specific gravity of the electrolyte any bearing on the battery capacity?

(c) What is the lower limit of specific gravity before the batteries must be refilled?

(d) About how often does the battery need to have the electrolyte renewed under service conditions?

3. (a) Why does the "normal renewal" solution have a specific gravity of 1250?

(b) Why is the normal strength of the electrolyte in the battry less than this or about 1200?

(c) Why does the "first filling" solution have a specific gravity of 1210?

(d) Does temperature have any effect on the specific gravity?

(e) If the specific gravity mentioned is measured at 15° centigrade, what is the corresponding Fahrenheit temperature?

4. (a) Assume the specifications for a renewal electrolyte is as follows:

Potassium hydroxide, 10.3 ounces per quart of solution.
Lithium hydroxide, 0.5 ounce per quart solution.
Specific gravity not less than 1250.
Temperature, 60° Fahrenheit.
Distilled water.

(b) How many quarts of electrolyte are needed to refill the battery?

(c) How would you mix the hydroxide with the distilled water?

(d) Does the hydroxide dissolve in cold water? Hot water?

(e) What is going to determine the accuracy of your proportions?

(f) Should the electrolyte be strained after it is mixed? Why?

(g) Can sodium hydroxide be used in place of potassium hydroxide?

(h) Will the sodium electrolyte have the same specific gravity?

5. (a) Should the battery be left standing empty any length of time? If not, why?

(b) Can the electrolyte be kept in stock already prepared?

(c) What kind of material should be used in constructing a container to hold the electrolyte? Why?

6. To what height should the electrolyte fill the cell?

7. (a) Does the state of charge have any bearing on the job?

(b) How can you insure complete discharge of the cells?

8. What should be done to the battery after it is refilled in order to prepare it for service?

9. (a) How does the electrolyte enter the batter into action?

(b) Why does the electrolyte have to be renewed?

JOB NO. 5.

Operate and Adjust a Motor Generator Set for Charging Truck and Tractor Storage Batteries.

Description:

A typical motor generator set for charging batteries consists of a 2200 volt, three phase, induction motor direct connected to a three-wire direct current generator and mounted on a common bed plate.

The induction motor has a starting compensator with overload and low voltage releases.

The direct current generator has a voltage range of 60-80 volts between either of the outside lines and 30-40 volts between either outside wire and the middle wire. A rheostat in the shunt field is used for voltage control, and the machine has a flat compound characteristic throughout the voltage range under a continuous load of 250 amperes. The unbalancing permissible is 35 per cent.

1. (a) What methods are available for changing alternating current to direct current for battery charging purposes?

(b) What is the objection to using a Tungar or Mercury Rectifier for charging large capacity batteries?

(c) How does a converter compare with a motor generator set for battery charging purposes?

2. Does the method of charging, whether constant current or constant voltage, affect the selection of battery charging equipment? If so, how?

3. (a) Is constant speed desirable in a battery charging machine?

(b) Is constant voltage desirable under all conditions of load?

(c) Will an induction motor give constant speed for all loads?

(d) Will a flat compound direct current generator give constant voltage for all loads?

(e) On this set is there any hand adjustment for obtaining constant speed at any load? Constant voltage? If so, what?

4. (a) If the charging rate is 45 amperes, how many 12-cell lead

truck batteries could be charged by this machine without overloading it?

(b) If the charging rate is 45 amperes, how many 21-cell Edison truck batteries could be charged by this machine without overloading it?

(c) Does the method of charging affect the number of cells that could be charged at one time?

5. (a) How can you put the set in operation?

(b) What is a starting compensator and why is one used?

(c) What would happen if the full line voltage were applied to the motor when starting the machine? Why?

(d) Why is an overload release placed on the A. C. supply circuit? A low voltage release?

(e) How are these releases constructed and why do they operate?

(f) About what horse power induction motor will be required for driving the direct current generator?

(g) Are there any adjustments on the overload and low voltage releases? If so, how are they used?

(h) Should any batteries be on the charging circuit while starting the machine? Why?

(i) What determines the speed the machine will attain when starting? Can this no load speed be changed?

6. (a) After the machine is started, what adjustments are necessary with regard to the voltage of the direct current generator before placing batteries on the circuit? How are the adjustments made?

(b) Is there any switch in the shunt field circuit?

(c) Why does changing the rheostat change the voltage of the generator?

(d) What is the purpose of the series winding on the generator?

(e) Is there any adjustment that can be made on the series winding if necessary? If so, how would it be made?

(f) Why isn't an automatic voltage regulator always specified for the direct current generator?

(g) How can you determine the positive terminal of the generator? Is this necessary? Is it always the same for an installed machine?

(h) If a direct current generator fails to build up, what remedies would you apply?

(i) If a generator had its polarity reversed, what remedies would you apply? What could cause polarity reversal?

7.(a) What limits the number of batteries you can charge with this machine?

(b) Do you need any overload protection on the charging circuit? If so, what kind and how would it be obtained?

(c) Do you need low voltage protection on the generator?

(d) Do you need reverse current protection on the generator? If so, is any apparatus available on the market?

(e) Would a short circuit on the charging circuit stop the set?

(f) Is it practicable to have the generator breakers with a time limit device to prevent conditions outlined in (e)?

8. (a) What instruments and switches are needed for the generator circuit?

(b) What instruments and switches are needed for the motor circuit?

(c) Do you need an ammeter on each charging circuit with both constant potential and constant current charging?

(d) Do you need a series resistance in each charging circuit? If so, how large a resistance?

9. (a) Could you operate the set with a short circuited coil on the generator? On the motor? What would happen?

(b) Could you operate the set with an open coil on either machine? What would happen?

(c) If one of the alternating current wires were reversed, what would happen to the machines?

(d) What would be the result of reversing the generator leads? The generator field?

(e) Can any emergency repair be made with conditions in (a) and (b) so as to operate the set? If so, what?

(f) How would you connect a watt-hour meter for measuring the direct current output of the set? The alternating current input?

(g) Why is 35 per cent unbalancing specified? How is the middle wire connected?

JOB NO. 6.

Operate and Adjust a Synchronous Converter used for Charging Truck and Tractor Batteries.

Description:

A typical synchronous converter for charging truck and tractor batteries consists of a machine with a three-phase alternating current side supplied through transformers with 10 tap secondaries from a 220-volt line.

The direct current side has a continuous rating of 250 amperes, with a voltage range of 60-80 by means of selector switches on the transformers. The voltage is adjusted for flat compounding under load by means of transformer reactance and a series winding. A three-wire direct-current system with 35 per cent unbalancing is arranged by tapping the third wire to the secondary neutral of the transformers supplying the A. C. side, the transformers being star connected. This gives a 30-40 volt circuit between either outside wire and the middle wire.

The A. C. supply is protected by overload and low voltage releases.

1. (a) Does the method of charging, constant current or constant potential, modify the specifications for a converter? If so, how?

(b) Why are transformers used on the A. C. side?

(c) Are the transformers of the auto type or of the two-coil type?

2. (a) How can the converter be started?

(b) Should the field be open or closed during the starting period?

(c) Should a low voltage be applied during the starting period? If so, why?

(d) Could the machine be started from either side?

(e) Should any batteries be on the direct current circuit during the starting period?

(f) How can proper polarity be obtained when starting from the A. C. side as an induction motor?

3. (a) Will the converter operate at a constant speed independent of the magnitude of the load? If so, why?

(b) Can the speed at which the converter operates be controlled by apparatus at the machine? Why?

(c) Does the direct current voltage remain constant for any load variation? If so, why?

(d) Why does changing switches on the transformers change the D. C. voltage?

(e) What effect has changing the shunt field current on voltage? Watts? Armature current? Why?

(f) Can you name any piece of apparatus that will change the voltage automatically and eliminate hand-controlled selector switches?

4. (a) How many 12-cell lead batteries will this machine charge?

(b) How many 21-cell Edison batteries?

(c) How many cells could be charged on the 30-40 volt circuit?

(d) Do you need any resistance in series with the charging circuits with either or both methods of charging? Why?

(e) What would happen if the machine were greatly overloaded?

5. (a) Why are low voltage and overload releases used on the A. C. side?

(b) What protective devices would you use in each charging circuit? In the main generator circuit?

(c) What switches are needed on the D. C. side?

6. What instruments are needed with the converter on the A. C. side? On the D. C. side?

7. Sketch connections and locate meters for the complete installation of this machine and its transformers.

JOB NO. 7.

Operate and Adjust a Constant Potential Charging Circuit for Charging Truck and Tractor Batteries.

Typical factory data on a truck battery:

12 cells.
225 ampere hours.
5 hour discharge.
1270–1280 specific gravity, charged.
1120–1130 specific gravity, discharged.
1.7 final cell voltage on discharge.

Charging apparatus is 15 Kw. D. C. three-wire system, with a voltage range of 60-80 between the outside wires and 30-40 between either outside and the middle wire. Thirty-five per cent unbalancing is permissable.

1. (a) What is the advantage of constant potential charging?

(b) What other method of charging is sometimes used?

(c) Does a battery have constant voltage during a charge?

(d) If a battery is charged on a constant potential circuit, what will happen to the value of the current as time elapses?

(e) What will be the extreme range of battery voltage?

(f) Why does the voltage of a cell continually change during a charge?

2. (a) Will the 30-40 volt circuit be of sufficient rating to charge the truck battery by the constant potential method if the charge started with the battery completely discharged?

(b) What will be the initial charging current through the battery if the supply voltage were 40?

(c) Do you have to know the internal resistance of the battery to compute the current taken on charge? Do you have to know the cell voltage? Why? Does it vary? Why?

(d) Is the internal resistance a constant during charge? Discharge?

(e) Assume the average internal resistance of a 225-ampere hour cell to be .0006 of an ohm, is the resistance greatest at the start of the charge or at the end of the charge?

(f) Is it allowable to assume some value of internal resistance to compute the initial current taken by the battery?

(g) What is the proper voltage per cell for most efficient charging by the constant potential method? Why?

3. (a) Would any external resistance be needed if the battery were charged from the 40-volt current? Why?

(b) Assume that .05 ohm is connected in series with the 40-volt circuit and that the internal resistance per cell is .0006, what would be the value of the initial current? Use any characteristic curve of a lead battery.

(c) Would this current be permissible?

(d) At the end of the charge assume that the internal resistance is .0008 ohms per cell and that the same external resistance is used, i. e., .05 ohm, what will be the approximate value of the current from the 40-volt circuit?

(e) Are ohmic resistance and voltage the only factors to be considered in computing charging currents?

(f) How can the electro-chemical effects best be considered?

(g) What is an adequate finishing rate current?

(h) Neglecting internal resistance entirely, what would be the initial and finishing rate currents if the battery is charged from the 40-volt circuit through .05 series resistance? Normal characteristic.

(i) Are these values permissible?

(j) How much voltage variation, during a charge, will there be at the battery due to line drop caused by the series resistance?

(k) What effect has the resistance of the connectors on the charging resistance needed? The variation in cell voltage?

4. (a) Assuming .1 ohm series resistance and two batteries charged in series from 80-volt circuit, what would be the initial and finishing rate current values if internal resistance is neglected?

(b) Are these values permissible?

(c) Compute the values of initial and finishing currents for a normal lead battery on assumed values of internal resistance, connector resistance, and the .1 ohm external resistance. Are these values permissible?

(d) Could the voltage variation of 10 volts on either circuit be used to make the .05 and .1 ohm series resistance adequate for their respective circuits?

(e) Will the drop in the lines be sufficient to change conditions if No. 4 wire is used and a maximum length of circuit is 60 feet?

(f) Compute the initial and final currents taken by two batteries in series, assuming values for internal resistance, connector resistance, line resistance and .1 ohm series resistance when charging from the 80-volt circuit? Assume characteristic curve for a normal lead battery.

(g) Would temperature affect your results in (f)? If so, how?

JOB NO. 8.

Determine the Charging Resistance to be Used for Charging a Philadelphia Storage Battery Co.'s Type 17 WML Truck Battery by the Constant Current Method.

The charging apparatus is a 15 Kw., 3-wire D. C. generator, with a voltage range of 60-80 between outside wires and 30-40 between either outside wires and middle wire and a permissible unbalancing of 35 per cent.

Manufacturers' Data:

240	ampere hours, normal rating.
5	hours discharge.
80°F.	temperature.
1270–1280	specific gravity, charged.
1120–1130	specific gravity, discharged.

1. (a) Will the question as to whether you use the 30-40 volt circuit or the 60-80 volt circuit affect your determination of charging resistance?

(b) Which circuit will permit of least resistance in the charging rheostat?

(c) Which circuit will operate at best efficiency?

(d) Could a rheostat figured for the 60-80 volt circuit be also used on the 30-40 volt circuit?

(e) Will the 30-40 volt circuit, if used alone, enable the battery to be charged at normal rate without exceeding the permissible unbalancing of the generator?

(f) What is meant by finishing rate and why is it used? What is its value?

2. (a) Assuming you have selected the 30-40 volt circuit, how are you going to determine the values of resistance you will need for all emergency conditions of charging the battery?

(b) Will four times normal rate be the maximum current you desire to put in the battery on charge? Will the instruments limit the current to be used?

(c) Is this value of current about the permissible limit of unbalancing with the generator used?

(d) Is 50 per cent of the finishing rate the minimum value of current that will be needed for charging this battery?

(e) Under what conditions will this value of current be desired?

(f) Will three values of resistance be determined by the maximum current, the normal current and the finishing rate current?

(g) Will other values of resistance be needed? Is it important to have an absolutely constant current at all times during charge?

(h) What is the objection to a large number of points on a charging rheostat? To a small number?

3. (a) How can you determine the value of resistance needed for any current, any number of cells, and any supply voltage?

(b) If the supply voltage is greater than the battery voltage, how can the excess voltage be used?

(c) How is the following formula obtained:

$$\text{Resistance} = \frac{\text{supply circuit voltage} - (2.5 \times \text{number of cells in series})}{\text{charging current}}$$

(d) Why is the figure 2.5 used?

4. (a) Assuming supply circuit at 40 volts, what should be the
resistance for normal charging current,
resistance for four times normal current,
resistance for finishing rate current, and
resistance for low rate charging current?

(b) Do the resistances represented in (a) represent the extreme values of resistance desired and can they be readily obtained in one rheostat?

(c) Aside from the numerical value of resistance, what factors enter into the design of a charging rheostat?

(d) Does the current capacity affect design? If so, how?

(e) Does the temperature effect on resistance affect result? If so, how?

(f) Is cast iron a good resistance material? Brass? Carbon? Steel? Copper ? All alloys?

5. (a) If a rheostat were designed with extreme values of resistance as computed in 4 (a), what current would flow into the battery if it has been discharged to 1.7 volts per cell and is then

placed on charge on the 40-volt circuit with the maximum resistance in series?

(b) Is (a) the extreme condition you would encounter with this equipment?

(c) How many intermediate resistance points would you place on the rheostat?

(d) What is the objection to using a rheostat that gets hot?

(e) Should a rheostat be mounted where there might be danger of igniting the battery gases?

(f) Can you cover the resistance material of a battery rheostat with porcelain, clay or other fireproof material and thus obviate danger of fire and oxidation of materials? If not, why?

6. (a) If the formula given in (3) is used to compute the resistance when the battery is connected to the 30-volt circuit, what are the extreme values of the resistance?

(b) What is the meaning of this result?

7. (a) If the formula given in (3) is used to compute the resistance for charging the battery from the 60-volt circuit, how do these values compare with those computed when charging from the 40-volt circuit?

(b) Are the internal resistances and connector resistances considered in the formula given in (3)?

JOB NO. 9.

Charge a Willard MLBA-17, 12-Cell Truck Battery by the Constant Current Method.

Manufacturers' Data:

225 ampere hours.
45 amperes.
5 hours discharge.
1270–1280 specific gravity, charged.
1120–1130 specific gravity, discharged.
80° F. normal temperature.

1. (a) How can you determine whether or not the battery needs charging?

(b) What precautions must be taken when using hydrometer readings as indicators of a battery's condition?

(c) How can you check the hydrometer's readings as regards the state of charge?

(d) What indication of charge is given by an open circuit voltage reading? Closed circuit reading?

2. (a) Assuming you have a 30-40 volt range in the supply for charging the battery, what apparatus will you need to charge the battery by the constant current method?

(b) Will you need any resistance in series with the battery?

(c) What kind of equipment is used for the series resistance?

(d) How constant must the current be kept during the charge?

(e) Do you need to read the temperature and specific gravity of all cells during the charge? Of only one cell?

(f) Should you inspect the cells for amount and specific gravity of the electrolyte before charging?

(g) Should you correct electrolyte conditions before charging? If so, how can you be sure of a true reading of specific gravity after water is added?

(h) Should the filling plugs be removed while charging? If so, why?

3. (a) Will you charge at 45 amperes throughout the charge? If not, why?

(b) Why is a so-called finishing rate used in constant current charging?

(c) How do you know when to put the battery on the finishing rate charge?

(d) What effect has gassing on the charging operations? Is it injurious to the battery if excessive?

(e) If you desire a finishing rate 40 per cent of the normal rate, will you need to put extra resistance in series on the 40-volt charging circuit? If so, what value?

(f) How can you make sure of having the positive side of the line connected to the positive side of the battery? Are they marked?

(g) What would happen to the battery if you had wrong polarity in charging?

4. (a) What will you do if the temperature rises above 110° F. during the charge? Why?

(b) How can you tell when the charge is complete? What values should be constant and for how long?

(c) If the battery is only partially discharged when placed on charge, will you charge it the normal length of time?

(d) What would be the effect of an overcharge?

(e) Should water be added to the cells while charging is going on? When charge is complete?

5. (a) At the end of the charge, is the battery ready for service with no further inspection or adjustment?

(b) Will the battery increase in capacity if left standing unused? If so, why?

(c) How long should a charged battery be left unused without recharging? Does this depend on partial discharge conditions?

(d) How would you treat the battery if it had overdischarged?

(e) How would you treat the battery if it would not take a charge?

(f) What would be indicated if the battery would not retain a charge any length of time?

JOB NO. 10.

Charge an Ironclad Exide MVY-15 Truck Battery by the Modified Constant Potential Method.

A 15-Kw. D. C. 3-wire circuit is available, with voltage between outside wires of 60-80 and between outside and middle wires of 30-40. A series resistance of 8.1 ohm is in the 60-80 volt circuit and 0.05 ohm in the 30-40 volt circuit.

Manufacturers' Data:

220.5 ampere hours.
4½ hour discharge.
49 amperes.
1280 specific gravity, charged.
1130 specific gravity, discharged.

1. (a) Before you put this battery on charge, is it necessary to know anything about its state of discharge?

(b) How will you determine its state of discharge?

2. (a) Before putting the battery on charge, is it necessary to remove filling plugs of all cells and examine height and specific gravity of electrolyte?

(b) If the height is correct, but the specific gravity is low, what is indicated?

(c) How will you remedy the condition?

(d) If the specific gravity is high and electrolyte low, how would you remedy conditions?

(e) Is it advisable to examine every cell as to height of electrolyte and as to specific gravity, or only one cell?

3. (a) Before connecting the battery to charging line, what other examinations or tests will you make on the battery?

(b) How can you determine the positive terminal of the battery?

(c) How can you determine the positive side of the line?

(d) Will you connect the positive side of the line to the positive of the battery?

(e) What would happen if you connected positive of battery and negative of the line?

(f) Would the battery be ruined? Injured? Charged?

4. (a) Is it advisable to connect the battery to the 30-40 volt charging circuit or to the 60-80 circuit? Why?

(b) In the constant potential method, do you need any resistance in series with the battery? How much? Why?

5. (a) What is going to limit the initial current on constant potential charging?

(b) The final value of the current?

(c) Is there any definite maximum for the allowable value of the initial current?

6. (a) Will you take any readings during charge? Why?

(b) What will readings of specific gravity show you? Of current? Of voltage?

(c) Will you read all cell values of voltage and specific gravity? If so, why?

(d) Which cell is usually used as a pilot cell?

7. (a) How long will the charge continue?

(b) How can you tell when to stop charging?

(c) Would the battery be injured if the charging continued?

(d) Would gassing continue?

(e) What would continuous gassing mean?

(f) How long should gravity and voltage remain constant before stopping charge?

8. (a) If the current near end of charge gets below 40 per cent normal, how could you expedite the charge?

(b) Is this advisable?

(c) Why is a smaller current used to finish the charge?

(d) What is the limiting factor in regard to the magnitude of this current?

9. (a) Will you leave the filling plugs in while charging?

(b) Should you add water to cells while charging? If so, why?

(c) Should you use a thermometer while charging? Why?

(d) What would you do if the battery exceeded a safe temperature?

(e) What is the maximum allowable temperature?

(f) What mechanical effect has expansion and contraction, caused by temperature changes, on a battery?

10. (a) At the end of the charge, what do you do to the battery to get it ready for service?

(b) Is it advisable to wipe off the battery top with ammonia cloth?

(c) Will the battery be ready for service if left standing on charge table in a charged condition for a day or two?

(d) How long should it stand charged without attention?

11. (a) What value of initial current would flow if the .05 ohm series resistance were removed from the 40-volt circuit? Is this excessive?

(b) How would you treat a sulphated battery on this charging circuit?

JOB NO. 11.

Charge an Ironclad Exide MVY-15, 12-Cell Truck Battery at Abnormal Rates.

Manufacturers' Data—Normal Operation:

220.5 ampere hours.
4½ hour discharge.
40 and 12 ampere charging rate.
1270–1280 specific gravity, fully charged.
80°F. temperature.

1. Can abnormal charging best be accomplished by the constant current or by the constant potential method?

2. Under what operating conditions may abnormal charges be desirable? What is meant by "boosting" charges?

3. (a) Under what conditions is a charge at low rate desirable?

(b) Will operating conditions ever demand a charge at less than normal rate? Will such a charge be demanded by the battery itself in order to keep it in condition?

4. (a) Does the rate of charging have any relation to the ampere-hour discharge capacity of the battery? Explain.

(b) If the rate of charge is doubled, is the ampere-hour discharge capacity of the battery increased, decreased, or does it remain about constant? Why?

5. (a) What are some objections to high charging rates?

(b) Does a high rate of charge cause heating?

(c) Is heating detrimental to battery operation? If so, why?

(d) What is the upper limit of temperature permissible on charge?

(e) Does a high rate of charge cause excessive gassing? If so, why?

(f) What is the objection to gassing in a battery?

(g) Are the gasses obnoxious, poisonous, explosive or visible?

(h) Does the gassing injure the plates due to the displacement of active materials? If so, when does this become objectionable?

(i) Does the iron clad type of plate aid this battery under high charging rates? If so, how?

(j) What is the maximum rate at which the battery can be charged without excessive gassing?

(k) What is the maximum rate at which the battery can be discharged without excessive gassing?

6. (a) Do the charging equipment and the connecting leads have to be increased in capacity when charging at high rates? If so, why?

(b) Have the equipment and leads sufficient capacity to give truck batteries a one-hour boost charge while the operator is eating his noon-hour lunch?

(c) What economic considerations enter into the answer to (b)?

7. (a) Does the internal resistance of the battery increase when high charging rates are used? If so, why?

(b) What would be the effect of an increase of internal resistance on heating? On battery voltage? On battery capacity? On charging equipment?

(c) If the cell maximum voltage is 2.4 on charge at normal rate and increases to 2.5 on charge at three times normal rate, what has been the increase in internal resistance? What other factors may account for the change in the voltage?

8. (a) What limits the rate you can use in charging a battery?

(b) Under what operating conditions are boosting charges advisable?

9. (a) The following formula is given to determine the safe charging rate:

$$\text{Amperes} = \frac{\text{ampere hours already discharged}}{1 + \text{hours available for boosting}}$$

Is this based on theory or practice?

(b) What would be the safe charging rate for this battery when half discharged and one hour is available for boosting?

(c) Is intermittent excessive rate charging better than continuous excessive rate charging? If so, why?

JOB NO. 12.

Charge an Edison A 6, 21-cell truck battery by the constant current method, using a 15-Kw. 3-wire D. C. generator having 60-80 volts between outside wires and 30-40 volts between either outside wire and the middle wire.

Manufacturers' Data:

225 ampere hours.
5 hour discharge.
7 hour charge.
45 amperes, normal.

1. (a) How can you determine the state of charge of the battery?

(b) Should you examine the height of the electrolyte before charging? If low, would you add water? Electrolyte? If so, how much?

(c) How can you tell the positive terminal from the negative?

2. (a) Will you use the 30-40 volt circuit or the 60-80 volt circuit if the charge is to be made at normal rate?

(b) If you use the 60-80 volt circuit, how much resistance must you put in series? If so, where will you get this resistance?

(c) If you use the 30-40 volt circuit, how much resistance will you use in series?

(d) What means will you use to determine when the battery is charged?

(e) Will you have the filling holes uncovered or covered during charge? Why?

(f) Will you use a finishing rate on the battery?

(g) What effect has temperature on charging rates?

3. (a) How long will the charge continue?

(b) How long should the charge continue after the voltage becomes constant?

(c) Will specific gravity readings aid you in determining the length of charge? If so, why?

(d) Does an overcharge injure the battery?

4. (a) Will gassing occur all during charge or only during a certain part of the time?

(b) Has the gas any distinctive odor?

(c) At the end of the charge should the height of the electrolyte be properly adjusted? If so, how?

(d) At the end of the charge is the battery ready for service? If so, why?

5. (a) If the battery has been over-discharged, can you put it back into shape? If so, how?

(b) If the battery will not take a charge, what is indicated?

(c) If the battery loses its charge rapidly, what is indicated?

(d) How can you tell when the battery needs refilling with electrolyte?

6. What is the effect of charging at low rates?

JOB NO. 13.

Charge a 21-cell A 6 Edison truck battery by modified constant potential method. A 15-Kw. 3-wire D. C. charging circuit is available, with voltage between outside wires of 60-80, from outside to middle of 30-40.

Manufacturers' Data:

225 ampere hours.
45 amperes discharge rate 5 hours.
45 amperes charge rate 7 hours.
1.2 average discharge volts per cell.
80° F. normal temperature.

1. Do you need all the manufacturers' data to do the job?

2. (a) What mechanical inspection of the battery will you make before placing it on charge?

(b) What will you do if you find corroded terminals on the battery?

(c) What will you do if you find the electrolyte too low?

(d) What will you do if the specific gravity is incorrect?

(e) Is specific gravity of any importance in Edison cells? If so, why?

3. (a) Before charging the battery, is it necessary to make a test to determine its state of charge?

(b) How can you determine the state of charge?

(c) Is it advisable to have filling caps opened or closed when charging?

(d) Will you add water while battery is charging? Why?

4. (a) How does constant potential charging differ from constant current charging, as applied to alkaline batteries? Which do you consider better?

5. (a) Should the battery be placed on the 30-40 volt circuit or on the 60-80 volt circuit?

(b) Does the allowable amount of current determine the circuit to be used or does the voltage of the battery?

(c) How much resistance will you put in series with the battery on the 30-40 volt circuit? Is any needed?

(d) Is any needed on the 60-80 volt circuit? Why?

6. (a) If resistance $= \frac{\text{supply voltage} - (1.7 \times \text{number of cells in series})}{\text{charging current}}$ what value of R is necessary to limit initial current to four times normal on the 40-volt circuit?

(b) Does the initial current magnitude affect the Edison battery as much as the lead?

(c) What effect has room temperature on battery capacity?

(d) Will intercell connector resistance be appreciable?

(e) If the charging circuit and panel has been used for charging a 12-cell lead battery, what changes must you make before charging a 21-cell alkaline battery?

7. (a) What readings will you take during the charge?

(b) How will you determine the end of the charge?

(c) How does gassing affect the operation of charging?

(d) Can you tell state of charge by voltage, by specific gravity or by temperature?

8. (a) If the finishing current is very low, how does this affect the capacity of Edison batteries?

(b) Are low rate charges satisfactory for Edison batteries?

(c) Are high rate charges detrimental?

(d) Can boosting charges be used to great advantage with Edison batteries?

(e) How long should the voltage remain constant before the charge is complete?

(f) Can you get quicker results if you increase the charge rate near the end of the charge? Why?

9. (a) At the end of the charge is the battery ready for service?

(b) If the electrolyte is low, how can you remedy the condition?

(c) Will the battery stand charged a day or two without losing an appreciable amount of its charge?

(d) Should anything be put on the terminals or on the top of the cells before placing the battery back in service? If so, what?

(e) How long could the battery be left unused on charging table after being charged and still be ready for service?

(f) If the battery were partially discharged and left standing unused, would its capacity be greatly lessened with time? Why?

JOB NO. 14.

Charge an Edison A 6, 21-Cell Truck Battery at Abnormal Rates.

Manufacturer's Data—Normal Rating:

225 ampere hours.
5 hours discharge.
7 hour charge.
1.2 average voltage per cell on discharge.
45 amperes normal current.
80° F. normal temperature.

1. (a) Under what operating conditions will it be necessary or advisable to charge the battery at excess rates?

(b) Does the battery itself ever need any excess rate charge to put it in better condition?

2. (a) What are limiting features as regards the amount of excess charging current?

(b) Is the mechanical construction of the battery such as to permit all possible charging currents?

(c) Has the battery such electrical and chemical characteristics as to permit excessive charging currents? Very low charging currents?

(d) How does temperature affect the mechanical characteristics and electrical characteristics during excess charging?

(e) Is the ampere hour capacity for discharge at normal rate affected if an excess charging rate is used? If so, how?

3. (a) Does a high charging current affect the specific gravity more than normal charging current?

(b) Does it affect the gassing of the battery?

4. (a) If the battery is charged at, say, one-quarter normal rate, will the discharge capacity be different from that obtained from a normal charge? If so, why? If charged at two times normal rate?

(b) What is the upper limit of temperature allowable in charging and why is this temperature stipulated?

5. (a) Why must a compromise rating be used on the battery if we assume that

   1. A maximum efficiency of 97 per cent can be obtained from a 3½-hour normal charge.

2. A maximum capacity of 325 ampere hours can be obtained from a 16-hour normal charge?

(b) Could a battery be designed to have maximum efficiency and maximum capacity for the same normal charge? If not, why?

6. (a) Why does the cell voltage increase from 1.5 to 1.95 during charge at high rates as compared to normal rate charges?

(b) Is the internal resistance of the battery constant during a charge?

(c) Does the internal resistance change with charge rates?

7. (a) Are the gases from the Edison battery obnoxious? Explosive? Emitted constantly during a charge?

8. How does temperature affect the life and efficiency of the battery and how does it depend on charging rate?

9. Is it better to charge continuously at excessive rates until complete charge is obtained than to charge intermittently at excessive rates until a complete charge is obtained? Why?

10. What is the approximate critical temperature for the battery?

JOB NO. 15.

A General Lead Batteries Co.'s Type BH-15, 12-cell truck battery, which has been in normal operation, is brought into a charging station. Determine its condition and its remaining amphere-hour capacity, if any for dscharge at normal rate.

Manufacturers' Data:

220 ampere hours.
4½ hours normal discharge rating.
80°F. normal temperature rating.
1270–1280 specific gravity, charged.
1120–1130 specific gravity, discharged.

1. (a) What daca regarding the state of discharge and condition of the battery could you obtain by inspecting the plates?

(b) Can you get the same information more readily and in better detail by making tests?

2. (a) Will any data about the battery as obtained from the manufacturer help you do the job? If so, what?

(b) Would this data warrant your omitting some tests?

3. (a) Should the case and containers be examined carefully before making a test?

(b) How would a leak be indicated?

(c) How could you find a cracked jar?

(d) How would you determine the height of the electrolyte in the cells?

(e) Will you add any water or electrolyte to the cells before making your tests?

4. (a) If your first test is to be made with battery on open circuit, what reading can you obtain on the battery and on the cells?

(b) Will a reading of battery terminal voltage help you do the job?

(c) Will a reading of the voltage of each cell help you do the job?

(d) Will a reading of the specific gravity of each cell help you do the job?

(e) Will a reading of the temperature of each cell help you do the job?

(f) What readings are essential to complete the job? Why?

5. (a) Do you find the temperature, specific gravity and voltage the same on all cells?

(b) Compute the average open circuit cell voltage by using the battery terminal voltage and the number of cells

6. (a) Assuming you wish to make other tests to do the job, what procedure will you follow and what readings will you take?

(b) If it is desired to discharge the battery at normal rate, determine the approximate resistance to connect across the battery terminals.

(c) Will you need an ammeter to measure the current on discharge?

(d) What precautions do you take to prevent injuring the instrument in case of excess current?

7. (a) Assuming you can take and tabulate the same readings as in (3) and in addition the current, would you do so?

(b) Are all the readings necessary for an accurate completion of the job?

(c) If a pilot cell is needed, which cell with respect to the positive terminal is generally used?

8. (a) Having taken the data, examine it for differences and results. Do the cell readings differ?

(b) Is the temperature important? Why?

(c) Do the cell voltages differ? Why?

(d) Do the cell specific gravities differ? Why?

(e) Does the sum of the cell voltages equal the battery terminal voltage?

(f) What permissible variation can occur in cell voltages? In cell specific gravities? In cell temperatures?

(g) What would a very high temperature reading on a cell indicate?

9. (a) From the data compute the remaining ampere hours of the battery for normal rate discharge. The kilowatt hours.

(b) Which is the more important from the operating standpoint?

(c) From the energy standpoint?

(d) From the financial standpoint?

10. (a) If the battery were tested on a hot day, say 100° F., would the reading be different than when tested at 70° F.? Why?

(b) Can you correct reading for temperature? If so, how?

(c) Would the capacity as determined by test at one temperature be the same as the capacity determined by tests at another temperature? Why?

11. (a) Put the battery on normal discharge and take readings to enable you to determine the actual ampere-hour and kilowatt-hour capacity of the battery; does the data check with your predicted results?

(b) Write specifications in detail for testing a battery to determine its condition at any time during normal operation; are these specifications applicable to any type of lead battery?

JOB NO. 16.

An A6, 21-cell Edison truck battery, which has been in normal operation, is brought into a charging station. Determine its condition and remaining ampere-hour capacity, if any, for discharge at normal rate.

Manufacturers' Data:

225 ampere hours.
45 ampere normal discharge rate.
5 hours normal discharge time.
80° F. normal temperature.

1. (a) What data and apparatus do you need to do the job?

(b) Will factory data help you do the job?

(c) Can any test data be eliminated by using the factory data?

(d) Has an Edison battery a name plate?

2. (a) What open circuit data can you obtain on the battery?

(b) Will temperature readings help you do the job?

(c) Will readings of voltage or of specific gravity?

(d) Will you take readings on each cell?

(e) Does a variation in voltage, temperature and specific gravity occur in cell readings? Why?

(f) What are permissible variations?

(g) Will you examine the height of electrolyte in each cell? If so, why?

3. (a) Is the open circuit data sufficient for you to do the job?

(b) Do you wish to take closed circuit data? What tests will you make and what readings will you take?

(c) If you desire to discharge the battery at normal rate, determine the approximate resistance to be connected across the battery.

(d) What precautions will you take to prevent injuring the ammeter?

4. What closed circuit readings will you take to do the job?

5. (a) Is it necessary to read voltage, temperature and specific gravity on each cell? Why?

(b) What cell is used for a pilot cell in ordinary testing?

(c) What differences are allowable in cell readings of volts, temperatures and specific gravities?

(d) Why do these differences occur, and how can they be remedied?

(e) Are they detrimental to normal operation?

6. (a) Can you compute the remaining ampere hours in the battery from the data taken?

(b) The kilowatt hours?

(c) How do they compare in value?

7. (a) If the battery were tested on a hot day, say 100° F., would the capacity be changed? How much? Why?

(b) If the battery were tested on a cold day, say 20° F., would the capacity be the same?

(c) Can you predict capacity at normal temperature from a test made at abnormal temperature conditions?

8. (a) How can you check your predicted capacity?

(b) How do your results check?

(c) Write specifications for testing an Edison A6 battery for condition and state of charge under normal operating conditions.

JOB NO. 17.

Put a Gould 17 V149 Truck Battery in Dry Storage.

Manufacturers' Data:

222.5 ampere hours.
1270–1280 specific gravity, charged.
1120–1130 specific gravity, discharged.
5 hour discharge.
40 and 16 amperes charging rate.

1. (a) When are batteries put in dry storage?

(b) Why is wet storage not better than dry storage for long storage periods?

2. (a) Should the battery be charged or discharged before being prepared for dry storage? Why?

(b) How long should the battery be charged?

(c) How long after gravity and voltage are constant should the charge be continued?

(d) What should be the readings of the voltages? Of the specific gravity?

(e) How high above the plates should the electrolyte be at the beginning of the charge?

3. (a) Will you keep a record of the specific gravity of the electrolyte in each cell at the end of the charge? If so, why?

(b) Will you preserve the electrolyte? Why?

(c) How long will you wait after the electrolyte is removed before adding water?

(d) May heating occur? Why?

(e) Is it essential to get all the electrolyte out of the cell before adding water?

(f) What kind of water should be added?

(g) How high above the plates should the water stand in the cells?

(h) If any electrolyte is spilled on the case, how will you remove it?

4. (a) How long should the battery stand when filled with water? Why?

(b) Would the battery keep in good condition if stored with only water in it?

(c) Should battery be charged or discharged when filled with water?

(d) Would charging or discharging when filled with water shorten the time taken in preparing for dry storage?

(e) Can the battery also be placed in dry storage by removing cells from case, then removing the elements and washing them several times in distilled water? If so, which is the better method?

5. How would you dismantle the battery after the water is emptied? (See Job No. 1).

6. (a) In what kind of a place should the plates be stored?

(b) Should they be put on top of each other?

(c) Can the separators be stored and used again? How? Is it advisable?

(d) Can the perforated rubber sheets be stored and used again?

(e) Can the jars be used again?

(f) Will all other parts of the battery be suitable for use after storage?

(g) Should the jars be stored dry?

7. (a) For wet storage purposes, compare the method of a monthly charge with that of placing the batteries on a so-called trickle charge as to efficiency and operating advantages.

(b) What minimum value of current should be used for trickle charge storage?

(c) What value of current should be used for the monthly charge method of wet storage?

JOB NO. 18.

A Gould 17 V149 truck battery has been in dry storage for 14 months. Install the battery in a truck for service purposes.

Manufacturers' Data:

222.5 ampere hours.
5 hour discharge.
1270–1280 specific gravity, charged.
1120–1130 specific gravity, discharges.
40 and 16 amperes, charging rate.

1. (a) In order to reassemble the battery, what new parts might be needed?

(b) Where can new parts, if desired, be obtained?

(c) Can they be kept in stock?

2. (a) What specific gravity electrolyte will you use for refilling the jars?

(b) What effect has temperature on the specific gravity desired?

(c) If no record is available of gravity before storage, what specific gravity can you use?

3. Assemble the battery. (See Job No. 2).

4. (a) How soon after filling the battery with electrolyte will you charge the battery?

(b) Why must time elapse?

(c) Has the temperature anything to do with the length of time the battery must stand before being charged?

5. (a) At what rate will the charge be made? Why?

(b) What would a boosting charge do to the battery?

(c) How long will a charge be continued? Why?

(d) In using constant potential charging, what precautions must be taken at start of charge?

(e) When using constant current, will the finishing rate be different from the initial rate? If so, why?

6. (a) What measurements will indicate how long the charge will be continued?

(b) If a new battery can be conditioned in 60 hours, why will an old one often take 75 hours?

(c) Why should temperature readings be taken during charge?

(d) What will you do if the temperature becomes greater than 110 degrees F. during the charge? Why?

7. (a) After charging is the battery ready for service?

(b) What should be the level of the electrolyte?

(c) Should the specific gravity be read before adding water or after? Why?

(d) How will you determine the real specific gravity after water has been added to the electrolyte? How long a time is necessary if you charge the battery?

(e) Will a reading of specific gravity after the battery has been discharged for a short time give as good an indication of specific gravity as a reading based on a charge of the same length of time?

(f) What will be the specific gravity when the battery is charged and properly filled?

8. If the temperature is different from normal, how can you correct readings of specific gravity?

9. (a) Why should the top and sides of battery be wiped off before putting the battery in service?

(b) Can you use ammonia on the cloth? Why?

(c) Is the battery ready for service?

(d) Should any coating be put on the exposed terminals of the battery?

JOB NO. 19.

Install a new General Lead Batteries Co.'s Type BH-15, 12-Cell Truck Battery.

Manufacturers' Data:

220 ampere hours.
4½ hour discharge.
1270–1280 specific gravity, charged.
1120–1130 specific gravity, discharged.
43 and 15 amperes, charging rate.

1. (a) What information does the name plate give you about the battery?

(b) Do you understand the meaning of all the symbols on the name plate?

2. (a) What are you going to do with the battery in order to have it ready for service?

3. (a) What procedure would you follow in unpacking the battery?

(b) Why should you unpack the battery in an upright position?

(c) When should covers to the filling plugs be removed?

4. (a) Why should all the packing material be removed and the top and sides of the battery wiped off and the whole unit carefully inspected?

(b) What would you do with the battery if a jar were broken?

(c) If a filling plug were broken?

(d) If a jar leaked?

(e) If the terminals were corroded?

5. After battery has been inspected for mechanical defects, what is the next item to inspect?

6. (a) Are batteries ever shipped unfilled? When?

(b) How high above the battery plates should be the electrolyte?

(c) If the electrolyte is low, will you add water immediately? If so, why?

(d) In case there is indication that the electrolyte has been spilled, would you add water?

(e) How can you determine whether evaporation or spilling has caused lowering in the height of electrolyte?

7. (a) What should a hydrometer read when placed in the cells?

(b) Will you take hydrometer readings before or after adding water or electrolyte, or both? Why?

(c) How can you insure getting a true reading of specific gravity after adding water?

8. (a) Is the battery shipped in a charged condition, so that it is ready for service?

(b) Does a battery decrease in capacity when standing charged, but unused? If so, how much and at what rate?

9. (a) What rate is best used for the conditioning charge?

(b) Should the filling plugs be in or out during the charge? Why?

(c) How long will the charge be continued?

(d) Where should specific gravity and voltage be measured?

(e) What indications as to voltage and specific gravity will show that the conditioning charge has been completed?

(f) How often should readings be taken?

(g) How long a time will the charge be continued after no appreciable change occurs in specific gravity or voltage?

10. (a) How can excessive gassing be prevented during the last part of the conditioning charge?

(b) What should be the finishing rate as compared to the normal rate?

(c) What effect will temperature have on making the conditioning charge?

(d) Should water ever be added while charging? Why?

(e) If the specific gravity is too high at the end of the charge, how will you bring it to the proper point?

(f) Why is water the only thing needed by a normal battery if no leakage has occurred?

11. (a) What will you do to remove any water or electrolyte spilled on top of the battery after conditioning charge?

(b) Is the battery now ready to be installed in a truck?

(c) Should anything such as vaseline be put on terminals?

12. How often and why should a battery be given an equalizing charge?

JOB NO. 20.

Determine the efficiencies and characteristics of a General Lead Batteries Co.'s Type BH-15, 12-cell lead battery under normal test conditions.

Manufacturers' Data:

220 ampere hours.
4½ hour discharge, normal.
1270–1280 specific gravity, charged.
1120–1130 specific gravity, discharged.
1.7 end point of voltage of cells for normal discharge.
43 and 15 amperes, charging rate.
80° F. normal temperature.

1. (a) Can the ampere hour and the watt hour efficiencies be determined by either the constant current method or the constant potential method of charging?

(b) What is meant by the term "efficiency?"

2. Which method lends itself most readily to test operations and to numerical computations?

3. What mechanical inspection of the battery is essential before testing it? How high above the plates should be the electrolyte?

4. What electrical and chemical tests should be made before determining the efficiency of the battery?

5. (a) Is there any reason for charging and discharging the battery several times at normal rates before taking data for an efficiency determination?

(b) Will such procedure insure average conditions, or best conditions, of the battery?

(c) What readings are you going to take to determine the efficiencies characteristics of the battery? How are efficiencies expressed?

(d) Is temperature of any importance in battery tests?

(e) Could a constant temperature be maintained? If so, how?

(f) Can readings be corrected for temperature changes? Should they be?

(g) What temperature variations should you find between beginning and end of charge and of discharge with outside temperature at about 70° F.?

6. (a) Will you first take data on charge or discharge?

(b) What advantages are there in starting with a fully charged battery and getting the discharge characteristics before getting the charge characteristics?

(c) Should you measure the open circuit voltage, the temperature, and the specific gravity of each cell or of one cell? Are these needed in an efficiency test?

(d) Does the voltage fall off rapidly at start of discharge? Why?

(e) What time elements should be used between readings at the start of the discharge?

(f) What time elements should be used between readings during the flat part of the discharge curve?

(g) Is it advisable to plot observed values to time as the discharge continues? Why?

(h) Do you desire the total time of discharge and the open circuit readings at the end of discharge?

(i) Will you discharge for 4½ hours or will you discharge until cell voltage drops to 1.7? Why?

(j) Should the amount of gassing be noted? Why? What does gassing signify?

7. (a) Will you start the charge immediately after discharge has ceased? What change would occur if time elapsed?

(b) What time elements between readings will you use during the different parts of the charge curve?

(c) Will you continue the charge at normal rate or will you use a finish rate in this efficiency test?

(d) Will you charge for seven hours or will you charge until the voltage and specific gravity remain constant?

8. (a) What would a curve of charge and discharge voltage plotted to time tell you about the battery?

(b) Compute the ampere-hour efficiency.

(c) Compute the watt-hour efficiency.

(d) Why does the discharge curve drop sharply at start of discharge and at the end of discharge?

(e) Why does the charge curve change its shape?

(f) What would curves of temperature and specific gravity plotted to time tell you about the battery?

(g) What effect has thickness of plates on shape of charge and discharge curves? Why?

9. (a) What is meant by "recovery" of a battery and why does it occur?

(b) If the battery were charged completely and then discharged for two hours at normal rate, and then charged one hour, and the operation repeated throughout a nine-hour day, what would be the approximate state of charge at the end of the day?

10. Explain the effect on efficiencies of charging at high and low rates? Of discharging at high and low rates?

11. (a) What is meant by "voltage efficiency"?

(b) What is the relation of the ampere-hour and voltage efficiency to the watt-hour efficiency?

(c) How do you estimate the minimum cost of charging from the output and the watt-hour efficiency?

JOB NO. 21.

Determine the efficiencies and characteristics of an A6, 21-cell Edison truck battery.

Manufacturers' Data:

225 ampere hours.
5 hour discharge.
7 hour charge.
45 amperes normal discharge.
45 amperes normal charge.
80° F. normal temperature.

1. (a) Should the battery be charged and discharged several times at normal temperatures before determining its characteristics?

(b) Will you take the battery out of service, charge it, and then have the efficiency determined from the discharge and the following charge?

(c) Do you desire to determine the best performance or the average performance of the battery?

(d) Is a determination of battery efficiency of any practical importance? If so, is the ampere-hour efficiency of more importance than the watt-hour efficiency?

2. (a) Does the constant current method lend itself better to the efficiency testing than the constant potential method? If so, why?

(b) Is it advisable to have the current constant during the whole of the charge, or should a finishing rate be used when making efficiency tests?

(c) In what order is it advisable to take the runs, i. e., start with battery charged or start with battery discharged? Why?

(d) Will voltage or will time determine the end of the discharge?

(e) What should be the cell voltage at the end of discharge?

3. (a) What data will you take to determine the efficiency and operating characteristics of the battery?

(b) Will readings of temperature, cell, specific gravity, cell voltage and battery voltage all be necessary?

(c) Will the temperature be constant during the test? If not, will your results be affected?

(d) Will you allow a temperature range corresponding to normal operating before any correction of readings will be made?

(e) Will you read specific gravity, volts and temperature on all cells, or on one cell?

(f) Should each cell be filled with electrolyte to normal level before starting the test?

4. (a) What time elements between readings will you use on discharge?

(b) What readings will you take at the same time?

(c) Why should the time elements between readings be short at the beginning and at the end of the run?

(d) Should all data be plotted in the form of curves? Why?

(e) Does the shape of the voltage curve depend on the specific gravity? On temperature? On internal resistance? If not, upon what does it depend?

5. (a) Should you start the charge immediately after the discharge?

(b) Should you use the same time elements between readings on charge you use on discharge?

(c) Should you read the same items on charge that you read during discharge?

(d) Should you notice when gassing starts and the relative amounts of gas emitted at different times during the charge? If so, why?

(e) Should your data be plotted on the same curve sheet as the discharge data? If so, why?

6. (a) What difference do you find between the ampere-hour efficiency and the watt-hour efficiency?

(b) What methods are available for determining these efficiencies from the curves?

(c) Is a calculation from average ordinates sufficiently accurate?

7. (a) Why does the voltage curve have a peculiar shape during charge?

(b) Why does the voltage curve on discharge assume a different shape than the charge curve?

(c) Why is the value of the voltage required for constant current charging constantly changing?

(d) Can you determine the internal resistance from the data taken?

(e) Does the time elapsing between end of discharge and beginning of charge affect your results? If so, why?

8. Would you get the same efficiencies if you charged at higher rate? A lower rate? Why? Explain.

9. Would you get the same efficiencies if you discharged at a higher rate? At a lower rate? Explain.

10. What is the voltage efficiency of the battery?

JOB NO. 22.

Burn connectors on a lead battery by the arc method and by the blow torch method.

1. (a) Do you consider burned connectors better than bolted connectors for truck batteries? If so, why?

(b) Are contact resistances and corrosion effects important items in battery operations?

(c) Under what conditions does it become necessary to burn connectors?

(d) Should a lead burning outfit be carried as part of the truck equipment?

(e) Could any type of lead burning outfit be improvised in an emergency? If so, what?

2. (a) What will you do to the connectors to prepare them for burning?

(b) What tool will be used to clean connectors?

(c) What effect does acid on the contacts have on the burning operation?

(d) Should the battery be protected from the flame by means of a damp cloth? By means of waste in the filling holes? Which is better?

(e) Under what conditions of burning should the gases be blown out of the cells?

(f) How can the cell gases be removed?

3. (a) How can you build up the battery posts if they are too short for burning?

(b) How can you tell the posts are too short?

(c) How ought a post, whose connector has been bored out, be treated before burning on a new connector?

(d) How ought a post that has had its connector removed by a puller be treated before burning on a connector?

(e) What tool can best be used for trimming posts?

4. (a) What equipment is used for arc burning?

(b) What is the objection to using a large carbon? A small carbon?

(c) How many cells do you need in series to operate the arc?

(d) Do you desire the arc or the hot carbon to do the burning?

(e) How far below the top of the connector is the top of the post for arc burning?

(f) Why should the top of the post be melted first in putting on a connector?

(g) Why is a rotary motion of the carbon and lead filling from the inside outward and upward advisable in arc burning?

(h) What would happen if the carbon remained stationary on the connector any length of time?

(i) What do you do to the top of the connector to get a finished job after arc burning is complete?

5. (a) In using an oxygen-illuminating gas burner, what kind of tip is advisable?

(b) In adjusting for burning, why is it advisable to use only four or five pounds' pressure of oxygen for the torch?

(c) What would likely happen if the tip became closed and oxygen was forced by higher pressure back into the illuminating gas mains?

(d) How can this be prevented?

(e) Why is it advisable to adjust the flame to about 8-inch length with the inner blue flame about 1 inch long for burning?

(f) Has the flame a characteristic hissing sound when properly adjusted?

(g) What part of the flame should be applied to the lead when burning?

(h) Why must the flame be used rapidly and removed frequently from the lead?

(i) Should the center of the post be first reduced, and then, using a rotary motion, build up the connector until flush with the top?

(j) Can a smooth top be obtained with the flame?

(k) What is meant by a hard flame? A soft flame? A reducing flame?

6. (a) Can a hot soldering iron be used for lead burning?

(b) Why is it inferior to either the arc or the torch?

(c) Could an ordinary gasoline glow torch be used for lead burning?

(d) Would an alcohol torch be suitable for lead burning?

JOB NO. 23.

A two-ton truck with rated load, equipped with a National Carbon Co.'s Type VV 1215Z, 12-cell battery, has its battery tested at a stopping point on a long trip and finds it reading as follows:

Voltage, 21.8.

Specific gravity, 1140.

The truck has still 10 miles to go, with no charging station intervening. At the point where the truck has stopped there is available for charging purposes a Signal Corps 2-Kw., 25 or 115 volt gas engine driven D. C. generator. GET THE TRUCK TO ITS DESTINATION.

Manufacturers' Data:

1270–1280 specific gravity, fully charged.
1120–1130 specific gravity, discharged.
220 ampere hours at normal discharge.
4½ hour discharge, normal.
35 and 14 amperes, charging rate.

1. Can the truck reach its destination under its own power without charging the battery?

2. Can the 2-Kw. set be utilized to charge the battery?

3. (a) How many amperes would be available if the 25-volt circuit of the generator were used?

(b) If the 115-volt circuit were used?

(c) How much overload can be obtained from the charging set?

4. (a) Will you use the 115-volt circuit to charge the battery?

(b) How long will it take to charge the battery from this circuit?

(c) Will you have to put some resistance in series with the battery? Why?

(d) What will you use for this resistance?

(e) If there were no question about generator capacity, would you need a series resistance?

(f) Will you charge at constant current or at constant potential, using the 115-volt circuit?

5. (a) Will you use the 25-volt circuit to charge the battery?

(b) How can you connect the battery to the circuit so it will be charged?

(c) If you have a lead burning outfit, will it be quicker to charge with the 115-volt or 25-volt circuit?

(d) Can you improvise a lead burning outfit if you cut a connector for charging from the 25-volt circuit?

(e) What connector will you cut and how will you connect the cells to the charging circuit?

6. (a) Will you use any series resistance if charging from the 25-volt circuit with your new connection of cells?

(b) Will you charge at constant current or constant potential?

7. (a) Under the conditions, what method of charging and which circuit of the generator will enable you to reach your destination in the shortest time?

8. Which method involves the maximum labor and trouble?

9. State conditions under which you would use the 25-volt circuit. The 115-volt circuit.

JOB NO. 24.

Answer some general questions about batteries and their operation.

1. (a) Do storage battery plates keep their shape and weight? Why?

(b) Why does not a form of electro-plating occur such as is noticed when zinc and copper are used in acid with electricity?

2. (a) If the material actually changes from lead to lead sulphate, why does not lead sulphate precipitate as sediment in the cell?

(b) Why do not trees and thick and thin places on the plates occur because of the reversible cycle?

3. (a) If the positive plate is lead or lead antimony alloy and contains lead peroxide, why is not the positive plate a storage battery of itself?

(b) Why does not it have a local charge and discharge of some magnitude?

(c) Why is not some other material than lead used as a grid to hold the lead peroxide?

(d) Should the grid be cleaned of sulphate?

(e) Should the positive grid have a great amount of surface exposed?

4. (a) If lead sulphate is the resultant of discharge action, why is a "sulphated" cell said to be in bad condition?

(b) Why should large sulphate crystals form on plates if a battery is left standing when discharged?

(c) Why cannot these large sulphate crystals be readily reduced by charging?

(d) Why is a long-continued charge at low rate used to reduce them rather than a short-time charge at high rate?

(e) Does sulphation help hold or cement the active materials together?

(f) What is the extreme of this action?

5. (a) Why does not a storage battery discharge itself when standing on open circuit fully charged? (Water can be decom-

posed into gases, when slightly acid, at a potential of only about 1.7 volts, yet the battery has a voltage of 2 and over).

(b) Why does it take 2.3 volts to cause a battery to gas on charge?

(c) Would impurities in the electrolyte affect the above conditions?

(d) In the active materials?

6. (a) Why do not the lead grids dissolve in the sulphuric acid?

(b) What kind of material forms on the outside of the lead?

(c) Why does it protect it from the acid?

(d) Why does not this protection also occur with respect to the spongy, active lead or the negative plate?

7. (a) Does a battery cool during discharge and heat during charge? If so, why?

(b) If it cools during discharge, what becomes of the energy taken from the air?

8. (a) Why is heat generated when sulphuric acid and water are mixed?

(b) Why does not the electrolyte in a battery always remain warm due to the above action?

9. (a) Why is the voltage of a cell a fixed thing approximately independent of the quantity of material or density of electrolyte?

(b) What exceptions are there to the above statements?

10. (a) Why does the capacity of a battery change with temperature?

(b) Does the capacity decrease with both extremes of temperature?

(c) What is the physical explanation of such action?

(d) What is meant by a "sluggish" battery?

(e) Why do some battery men short circuit an Edison battery for an instant just before using?

(f) Is there any truth in their reasons?

11. (a) What causes buckling of plates?

(b) Is buckling apt to occur in a modern vehicle battery?

(c) What limits the current output of a battery?

(d) Will a short circuit necessarily injure a lead battery? An alkaline battery?

(e) What differences in behavior is there between a thin plate battery and a thick plate battery?

(f) Where are thin plate batteries used?

12. (a) What is the effect of partial discharges on the capacity of a battery?

(b) What should be done to a battery that has been operating under conditions of partial discharge?

JOB NO. 25.

Determine conditions under which a truck equipped with an A 6, 21-cell Edison battery or one equipped with an Electric Storage Battery Co.'s Type MVY-15, 12-cell, will be ordered into service. Motor rating on truck is 45 amperes at 24 volts. Truck guarantee is four miles per hour on level with 4000-pound load. The battery is to operate the truck for a total of four hours actual running time per charge per day.

Manufacturers' Data:

Edison—A 6, rating, 45 amperes on 5-hour discharge.

E. S. B. Co.'s—MVY 15, rating, 49 amperes on 4½-hour discharge.

1. (a) Assuming identical trucks and their batteries fully charged, a load of 4000 pounds is to be carried from the station a distance of eight miles. The outside temperature is 98° F., which truck would you order into service?

(b) Would either truck do the job?

(c) Assuming a level road, which would do the job in less time?

(d) Under the same conditions, excepting an outside temperature of 18° F., which truck would you order into service?

(e) Would either truck do the job?

(f) Which truck would do the job in less time on a level road?

(g) In case you have time, could you prepare the batteries to do the job in better shape on a hot day? On a cold day?

(h) What would you do to the Edison? What would you do to the lead?

(i) Would these precautionary measures enable either or both the trucks to perform their four-hour guaranteed service?

2. (a) A load of 4000 pounds is to be taken a distance of 24 miles in ten hours. A charging station is located 12 miles from the starting point.

(b) Which truck would you order into service, assuming normal temperature and level road? Why?

(c) Would either truck do the job?

(d) Which would do the job in the shortest time?

(e) What maximum charging time could you allow on the lead at the half-way station? On the Edison?

(f) What maximum boosting charge would you utilize on the lead? On the Edison?

(g) Assume the road has such a grade that a speed of only three miles per hour can be made. Would this change your choice of truck to be ordered into service? How?

(h) If the road is steep and muddy so that you can average only two miles per hour, which truck would you order into service? Why?

3. (a) On a busy day, normal temperature, it is desired to work trucks under conditions such that only half-hour boosting charges are available every three hours. Which truck could do its guaranteed service for a 12-hour day? Why?

(b) Could each do it?

(c) Which would give service a longer time? Why?

(d) Would such service injure the Edison? The lead?

4. Which truck will have the greatest speed running light on a level road at normal temperature? Why?

5. Which truck will have the greatest speed on a level road under heavy load conditions at normal temperature?

6. A four-ton truck weighing complete with load 8500 pounds has become stuck in the mud. In the emergency, which truck would you order out for helping the stalled truck? Why?

Spotted something that isn't quite right?

Let us know and we'll fix it for future reprints.
And as a thank you for your diligence, we'll
send you a free Sothis book of your choice.

publisher@sothis.press

This page intentionally left blank

Made in the USA
Monee, IL
08 November 2026

67812140R00208